Real Relativity

Paul Sommers

2018

Preface

This book was first written in 1981 when the author was living aboard a 31-foot trimaran sailboat named SPACETIME. It was typed on a classic Underwood typewriter, which was cast overboard into the deep waters of the Atlantic upon completion. The figures were sketched by hand.

In 1998, the book was transcribed almost verbatim into an electronic latex file, and the figures were recreated electronically and stored as postscript files.

Einstein's theory of relativity was formulated in the early 1900's. The detection of gravitational waves in 2016 and the astrophysical importance of black holes have further validated the theory since 1981, but the theory itself has not changed.

This book is meant to convey the true essence of relativity. The revolutionary insights of Einstein cannot be properly understood by formulas and words about time dilation and Lorentz contraction. Einstein discovered that the universe is not a 3-dimensional Euclidean stage inside of which things act on each other while a mysterious time parameter increases. Instead, space and time are welded together in a unified 4-dimensional geometry. The geometry of Minkowski spacetime is essential for understanding the physics of the real world.

The presentation relies on no tools of calculus or trigonometry. It assumes only that readers are comfortable with algebra. They may be young students or mature professionals. It is hoped that some will be inspired to pursue the mathematics of differential geometry and progress to the forefront of theoretical physics where the challenge remains to reconcile the geometry of general relativity with the established phenomena of quantum theory.

Contents

Preface		i
0	INTRODUCTION	1
1	SPACETIME	5
2	MINKOWSKI SPACETIME	13
3	THE RELATIVITY OF VELOCITY	29
4	RADAR MEASUREMENTS AND SIMULTANEITY	31
5	THE ONE AND ONLY SPEED OF LIGHT	53
6	SWITCHING REFERENCE SYSTEMS AND "ADDING" VELOCITIES	59
7	TIME DILATION	77
8	THE TWIN EFFECT	83
9	LORENTZ CONTRACTION	99
10	GENERAL RELATIVITY	109
About the Author		121

Chapter 0

INTRODUCTION

The real world is not exactly the way it seems. Albert Einstein deduced that our human intuition about space and time is amiss. He was the first Earthling to discern the actual geometry of the 4-dimensional spacetime in which we all live.

Our failure to perceive directly this spacetime geometry is due to the very high speed of light compared with the speeds of everyday objects and creatures. It is amusing to imagine a universe which is set up just like our actual universe, except that light does not travel much faster than other common things. A salesperson who drives around while working would come home for lunch and find that his stay-at-home wife had already eaten supper and gone to bed. His wristwatch would have been running far slower than the clock on the mantle because he had been traveling at nearly the speed of light. A Cadillac driving past the house would be much shorter than when it was parked. In fact, all moving objects would be appreciably shortened. If you grew up in that hypothetical universe you would undoubtedly have reliable intuition about spacetime geometry.

We don't perceive those relativistic effects in our everyday lives because such changes in clock rates and changes in length are minuscule unless things are moving at speeds comparable to the speed of light, and the actual speed of light is incredibly fast compared to objects that we experience. And yet those relativistic effects are real. They are genuine features of our universe. Their reality has been well established by sensitive experiments, and the effects are very large for elementary particles which have been accelerated almost to the speed of light. These observed effects verify that the geometry of Nature is not that of our intuition – it is the spacetime geometry which Einstein discovered.

Fortunately, it is not too difficult for us to acquire an understanding of spacetime geometry, even though most of us grow up with very little intuition for it. That's the purpose of this book – to introduce you to the geometry of the real world. You won't need any fancy mathematics, though I hope it's OK with you if I use some basic algebra as one way to communicate. Please read leisurely, and examine each diagram as you come upon it. Getting the appropriate pictures in your mind is the essence of understanding any type of geometry. Spacetime geometry is no exception.

I think you will find it rewarding to learn about relativity and some of its wonderful implications. It is a mind-expanding endeavor which will give you a fresh perspective of reality.

I'm not going to burden you with a lot of evidence confirming the correctness of relativity. Einstein's discovery occurred early in the 20th century, and it has been well established by experiments and observations. Nor will I attempt to trace the history of ideas which led up to Einstein's discovery, even though the elegance of relativity is best appreciated when the discovery is seen in a historical perspective. You see, at the beginning of the 20th century physicists were struggling with a number of complicated and perplexing puzzles which had arisen from their experiments and theories. Relativity provided a very tidy solution to those complicated puzzles. It was only necessary for people to give up their old ideas about space and time and to accept, as real, the less intuitive spacetime geometry of relativity. I want to focus on this essence of relativity – the geometry of spacetime.

> *Note*: As you read, you will encounter a number of notes, remarks, calculations, and recreations set off in boxes like this. Like tributaries to a river, they are not the main flow of ideas, and yet they contribute to the strength of the main flow. I hope you will explore the boxes as you go. However, you might enjoy making some of those excursions another time, just cruising down the main stream for now.

> *Remark*: I'm not writing especially for physics enthusiasts, so I'll avoid using concepts like "energy," "momentum," and "inertial mass." Accordingly, this box will be the only reference to Einstein's most famous equation:
>
> $$E = mc^2$$
>
> Here E stands for the total energy of an object, m denotes its inertial mass, and c is the speed of light. The equation quantifies the idea that energy and mass are not really different. One can be converted to the other. Nuclear energy is the result of an atomic nucleus converting some of its mass into usable (or destructive) energy.
>
> Changing from old fashioned ideas of space and time to the geometry of spacetime required some changes in what physicists meant by "energy," "momentum," "inertial mass," etc. The equation $E = mc^2$ is part of that story.

Chapter 1

SPACETIME

If you want to understand what relativity is all about, the foremost concept you'll need is that of *spacetime*. I'd therefore like to introduce this concept with some deliberateness. First of all, we should have a common understanding of what is meant by the "dimension" of a space.

A railroad train moves along a one-dimensional space which is the railroad track. A train's position in that space can be specified by a single number which tells how far the train is from one end of the railway line. A ship moves on the 2-dimensional space of an ocean's surface. To specify a ship's position, two coordinates are necessary: its longitude and its latitude. Aircraft move in a 3-dimensional space which can be coordinatized by altitude, longitude, and latitude. The dimension of a space is equal to the number of coordinates which are necessary to specify a location in the space.

We humans generally think of ourselves as living in a 3-dimensional space. It is customary to use coordinates x, y, z to label the points in that space. Figure 1.1 illustrates how this three dimensional grid works. The point P in this picture is the point with coordinates (5,6,2), which is to say, it is the point (x, y, z) for which $x = 5, y = 6$ and $z = 2$. The point O has coordinates (0,0,0).

Let's say that the distance $\mathcal{D}$ between a point (x_1, y_1, z_1) and another point with coordinates (x_2, y_2, z_2) is given by this formula:

$$\mathcal{D} = \sqrt{(x_2 - x_1)^2 + (y_2 - y_1)^2 + (z_2 - z_1)^2} \qquad (1.1)$$

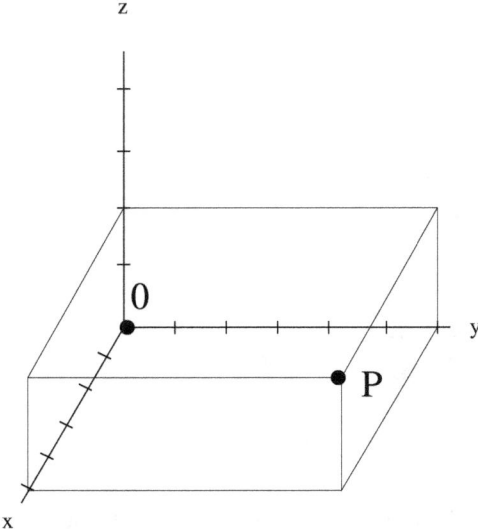

Figure 1.1:

> *Example*: In figure 1.2, the distance between (5,6,2) and (1,2,4) is
>
> $$\mathcal{D} = \sqrt{(1-5)^2 + (2-6)^2 + (4-2)^2} = 6$$

The 3-dimensional space in which distances are computed by this formula 1.1 is known as *Euclidean space*. It is the 3-dimensional geometry which is commonly studied. Just for comparison later on, let me mention that formula 1.1 can be written in a simpler-looking form by squaring both sides:

$$\mathcal{D}^2 = (x_2 - x_1)^2 + (y_2 - y_1)^2 + (z_2 - z_1)^2 \tag{1.2}$$

I'd like now to simplify my sketches of 3-dimensional space. Instead of showing a point P in space as in figure 1.3, I'll use the simpler picture of figure 1.4. The entire 3-dimensional space is represented by something that looks like a sheet of paper. Please regard this new picture as representing the full 3-dimensional space even though it doesn't look like it. It may help to think of it as a snapshot of 3-dimensional space.

Suppose now that P is the position of a golf ball as it is putted toward a hole at time $t = 0$. One second later the ball is at a new position Q. We can sketch both positions in one picture (figure 1.5). In fact we can exhibit the

Spacetime

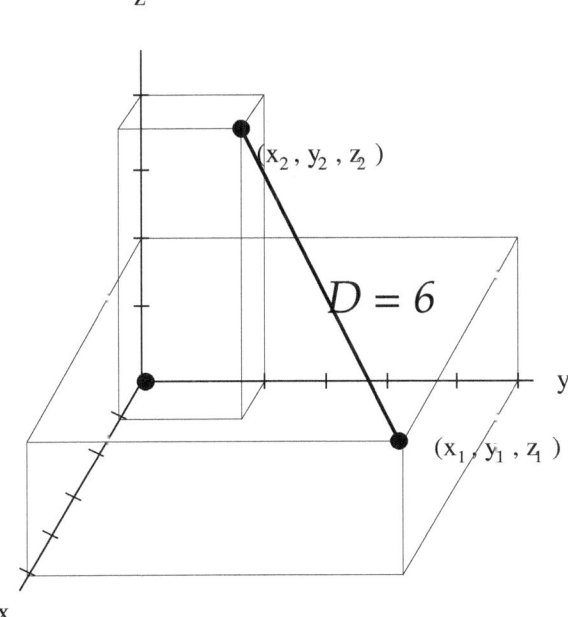

Figure 1.2:

Figure 1.3:

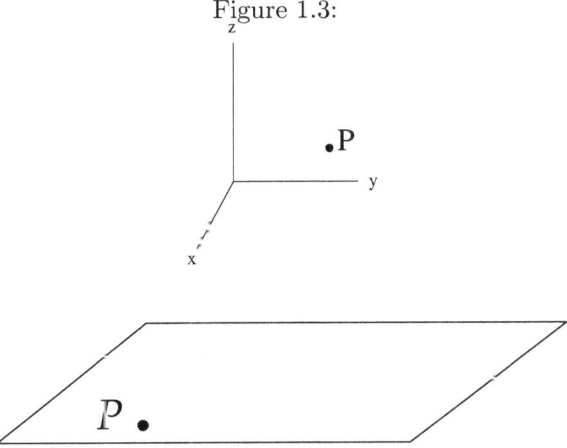

Figure 1.4:

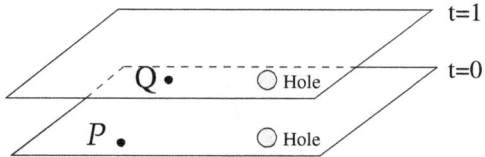

Figure 1.5:

golf ball's position for many seconds after striking, all in the same picture (figure 1.6). You can regard any one slice in this picture as a snapshot of

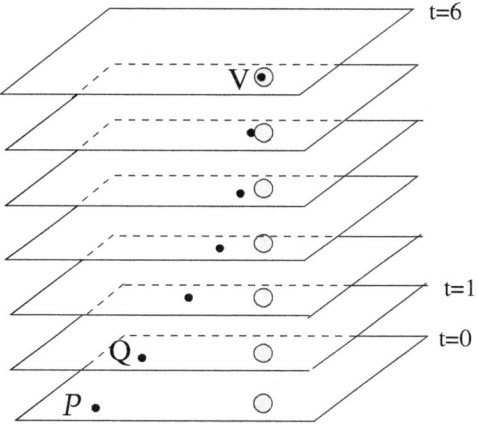

Figure 1.6:

3-dimensional space taken at a particular time. All together, the snapshots would constitute a motion picture of the putt. I have combined the whole time sequence into a single sketch which shows the position of the ball at seven different instants. We can do better. We can display the position of the ball at *every* instant between $t = 0$ and $t = 6$ (figure 1.7). The sequence of ball positions now appears as a curve in the sketch. Any one horizontal slice in the sketch still represents all of space at a single instant of time, and time increases as you move upward in the stack of snapshots.

This figure 1.7 depicts 3-dimensional space for some continuous period of time; it is called a *spacetime diagram*. Spacetime is 4-dimensional because four coordinates (x, y, z, t) are required to specify any point of it. The t-coordinate of a point specifies which slice it is on, and the x, y, and z coordinates locate the point in that 3-dimensional slice. A point in spacetime is called an *event* because it is associated with a particular time as well as

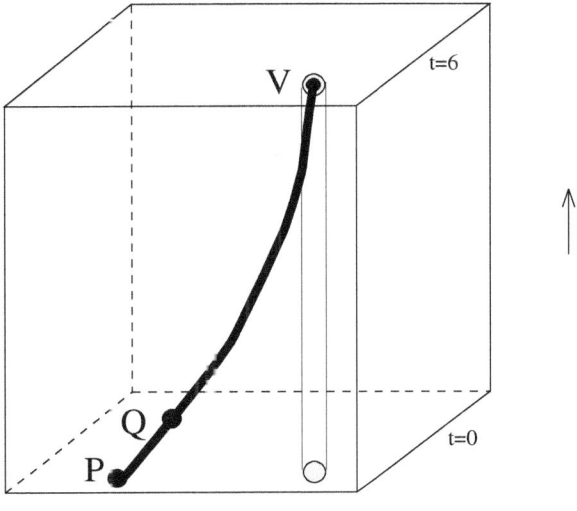

Figure 1.7:

a space location. The curve of events in figure 1.7 is called the *worldline* of the golf ball.

For the most part, we will be considering motion along only one spatial dimension. In the example above, we might assume that the golf ball moves along the straight line in space which joins its initial position to the hole. If, in another context, we contemplate a rocket flight to the Andromeda galaxy, we only need to consider motion along the straight line between ourselves and Andromeda. Whenever the relevant portion of space is only one-dimensional, it is superfluous to show snapshots of 3-dimensional space. We need only show the relevant one-dimensional space, sketched as a line and coordinatized by x, as in figure 1.8. The golf putt then appears as in

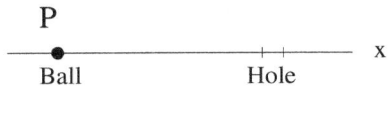

Figure 1.8:

figure 1.9. Or we can display the putt as in figure 1.10, showing the position of the ball at every instant of time during the putt. Prior to time $t = 0$ the ball is at rest. At event P it is struck. At event V its space position coincides with that of the hole, and it remains there as time continues.

In these last diagrams spacetime looks two-dimensional, the events being

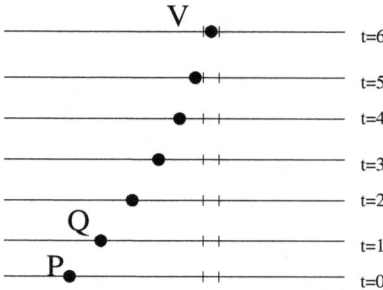

Figure 1.9:

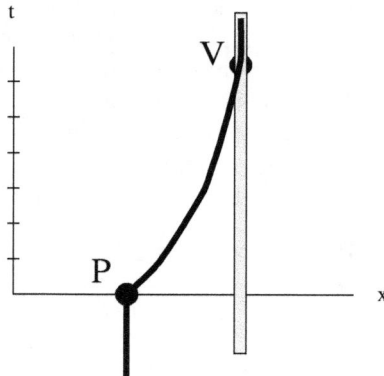

Figure 1.10:

labeled by the coordinates x and t. I have ignored the other two spatial dimensions simply because everything of interest is happening in the xt-plane. In the succeeding chapters, we'll continue to focus on the xt-plane of spacetime. That's where the interesting stuff is going on.

Chapter 2

MINKOWSKI SPACETIME

Indulge me, if you will, by looking at this diagram which shows two events in spacetime. Event E has coordinates (3,1), i.e. $x = 3$ and $t = 1$. Event

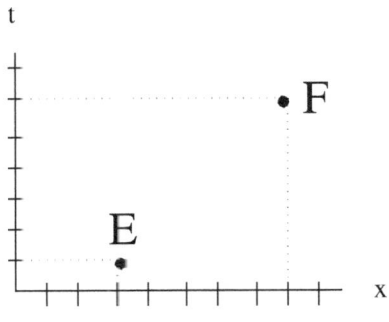

Figure 2.1:

F is (7,6). Suppose I were to ask you, "What is the separation of events E and F?" Chances are you would say, "Well, they are separated by 4 units in space and by 5 units in time." That answer gives *two* separations for the pair of events. I'd like you to consider Einstein's answer,[1] which gives a single number for the separation $\overline{EF}$ between events E and F:

$$\overline{EF} = \sqrt{-(7-3)^2 + (6-1)^2} = \sqrt{-16 + 25} = 3 \qquad (2.1)$$

---
[1] Einstein, A., Ann. d. Phys. 17, 891 (1905).

You see, in the previous chapter I introduced spacetime as a bunch of events, a three-dimensional space for each instant of time. But I didn't tell you how to compute distances or "separations" between events in spacetime. That's like giving you a balloon which you can push and pull into different shapes: the rubber surface is a certain bunch of points, but the distances between points of the balloon are not fixed. Spacetime is not pliable like a balloon; any pair of events has a fixed distance or separation between them. This is not simply a separation in space together with a separation in time, which would be two separations for a single pair of events. Space and time are welded together more intimately, and Einstein was the first person to be aware of this structure of spacetime.

Instead of writing a formula for the distance or separation between a pair of events, I'll first display the formula for the *interval* between events. Suppose E is some event with coordinates (x_1, t_1) and F is a second event with coordinates (x_2, t_2). The interval between them is given by

---

The Minkowski Interval Formula:

$$\mathcal{I}_{EF} = -(x_2 - x_1)^2 + (t_2 - t_1)^2 \qquad (2.2)$$

Here $\mathcal{I}_{EF}$ stands for "the interval between events E and F."

Remark: We can compute the interval between *any* pair of events, not just pairs lying in the xt-plane. If G is event (x_1, y_1, z_1, t_1) and H is (x_2, y_2, z_2, t_2), then the interval $\mathcal{I}_{GH}$ is given by

The Full Minkowski Interval Formula:

$$\mathcal{I}_{GH} = -(x_2 - x_1)^2 - (y_2 - y_1)^2 - (z_2 - z_1)^2 + (t_2 - t_1)^2 \qquad (2.3)$$

These formulae are reminiscent of the formula

$$\mathcal{D}^2 = (x_2 - x_1)^2 + (y_2 - y_1)^2 + (z_2 - z_1)^2 \qquad (2.4)$$

for the squared distance between a pair of points in Euclidean space. In the Minkowski interval formulae, however, time is included and there are some peculiar-looking minus signs. Notice that, unlike $\mathcal{D}^2$, the interval $\mathcal{I}_{EF}$ is not necessarily a positive number. Depending on the pair of events, it

Minkowski Spacetime

may be zero, positive, or a negative number. Let's fix our attention on one event E with coordinates (x_1, t_1). No matter how we choose a second event F with coordinates (x_2, t_2), we can compute the interval $\mathcal{I}_{EF}$. Let's look closer at the three possibilities:

1. $\mathcal{I}_{EF} = 0$. In this case events E and F are said to be *null-separated*. By examining the Minkowski interval formula, you can see that this happens for events (x_2, t_2) such that $|t_2 - t_1| = |x_2 - x_1|$. Another way to say the same thing is this: Event F is displaced from E as much in the t-direction as it is in the x-direction. All the events which are null-separated from E constitute the two lines through E which make a 45° angle with the coordinate axes. See figure 2.2. As a matter of

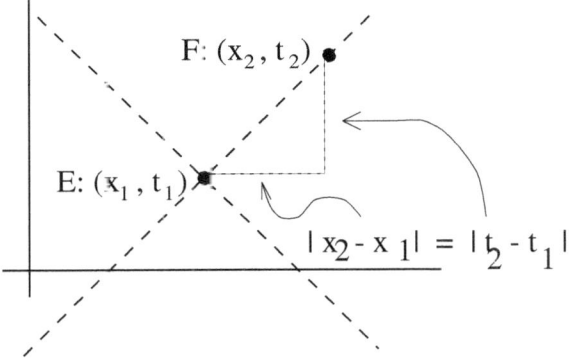

Figure 2.2:

notation, I will write

$$\overline{EF} = 0 \text{ for null-separated events.}$$

2. $\mathcal{I}_{EF} > 0$. Another look at the Minkowski interval formula will convince you that this happens if $|t_2 - t_1| > |x_2 - x_1|$, that is, provided F is displaced from E more in the t-direction than it is in the x-direction. Events E and F are said to be *timelike-separated* in this case. Events which are timelike-separated from E lie in the shaded region of the diagram below (figure 2.3). If E and F are timelike-

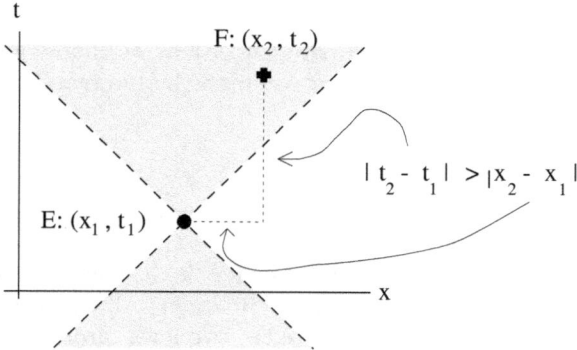

Figure 2.3:

separated, the square root of the interval is called the separation of E and F. I shall denote the separation by $\overline{EF}$, so

$$\overline{EF} = \sqrt{\mathcal{I}_{EF}} \text{ for timelike-separated events.}$$

3. $\mathcal{I}_{EF} < 0$. This happens when $|x_2 - x_1| > |t_2 - t_1|$, that is, when F is displaced from E more in the x-direction than in the t-direction. Events E and F are said to be *spacelike-separated* in this case. The shaded portion of figure 2.4 represents the events which are spacelike-separated from event E. Since $\mathcal{I}_{EF} < 0$ for spacelike-separated events, the square root of the interval $\mathcal{I}_{EF}$ is not a real number. It is therefore awkward to talk about the "separation" of such events. Instead, I will define the "distance" between E and F, in this case, as the square root of the positive number $-\mathcal{I}_{EF}$. I'll denote the distance by $\overline{EF}$, so

$$\overline{EF} = \sqrt{-\mathcal{I}_{EF}} \text{ for spacelike-separated events.}$$

For any pair of events E and F, you now know how to compute the number $\overline{EF}$. You first compute the interval $\mathcal{I}_{EF}$ using the Minkowski interval formula. If the interval is zero, then the events are null-separated

Minkowski Spacetime

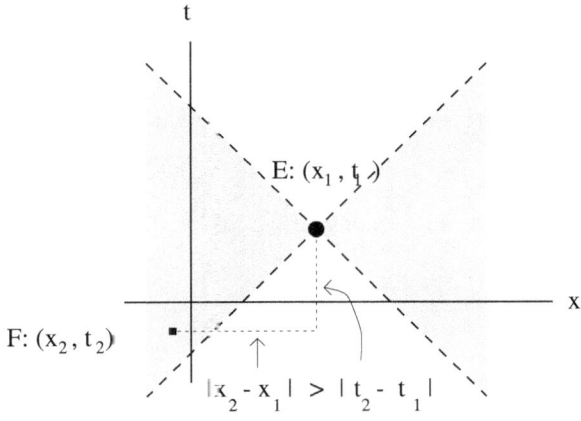

Figure 2.4:

and $\overline{EF} = 0$. If the interval is positive, the events are timelike-separated and $\overline{EF} = \sqrt{\mathcal{I}_{EF}}$. If the interval is a negative number, the events are spacelike-separated and $\overline{EF} = \sqrt{-\mathcal{I}_{EF}}$.

> *Note*: The formula $\overline{EF} = \sqrt{|\mathcal{I}_{EF}|}$ is valid in all three cases. You may wonder why I don't forget about the Minkowski interval formula and just write
>
> $$\overline{EF} = \sqrt{|-(x_2 - x_1)^2 + (t_2 - t_1)^2|} \qquad (2.5)$$
>
> Well, if you use this formula, you don't know whether $\overline{EF}$ represents a timelike separation or a spacelike distance. By taking the absolute value of the interval, that important information is lost. This formula for the quantity $\overline{EF}$ is fine, though, if you already know that E and F are spacelike-separated, for example.

Recreation: Consider the four events whose coordinates are given in figure 2.5. Show that

1. O and E are spacelike-separated with $\overline{OE} = 2\sqrt{2}$
2. O and F are spacelike-separated with $\overline{OF} = 2\sqrt{2}$
3. O and G are null-separated, so $\overline{OG} = 0$
4. E and F are timelike-separated with $\overline{EF} = 2$
5. E and G are timelike-separated with $\overline{EG} = 2\sqrt{2}$
6. F and G are null-separated, so $\overline{FG} = 0$

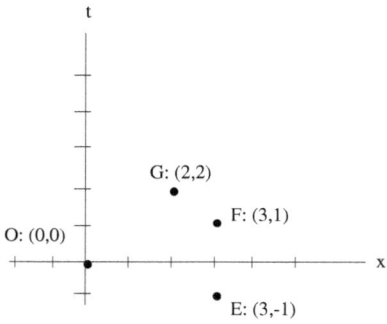

Figure 2.5:

More recreation: Consider any slice of spacetime obtained by specifying a fixed value for t. Show that the Minkowski distance between any pair of events in this slice is identical to the Euclidean distance of formula 1.1.

We have now got together all the ingredients of *Minkowski spacetime*.[2] The events of Minkowski spacetime can be labeled by coordinates (x, y, z, t), with every such set of four numbers corresponding to a single spacetime event. The interval between any pair of events is given by the full Minkowski interval formula 1.2. We will be investigating Minkowski spacetime in some

---

[2]Minkowski, H., Phys. Zeitschr. 10, 104 (1909).

detail as we study how it fits the real world.

When you see a diagram of the xt-plane of Minkowski spacetime, you must suppress your ordinary (Euclidean) notions about distances, and use the Minkowski interval formula instead. For example, in the left diagram of figure 2.6, the distance from event O to event E is *greater* than the distance from O to F. Similarly, in the right diagram, the separation of O and G is

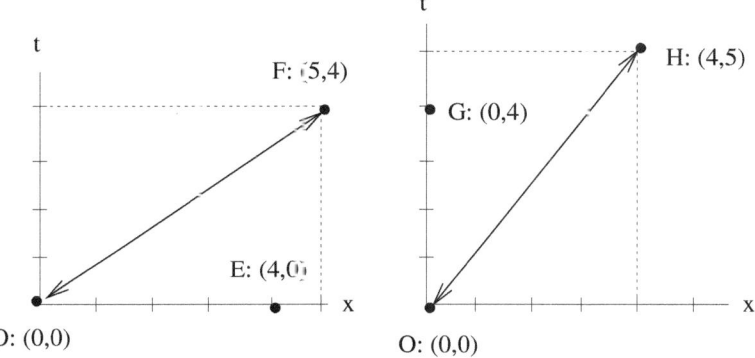

Figure 2.6:

greater than the separation of O and H.

In an ordinary Euclidean plane, coordinatized by x and y, the points which are at a fixed distance r from the origin make up a circle (figure 2.7). To say that a point with coordinates (x, y) is a (Euclidean) distance r from

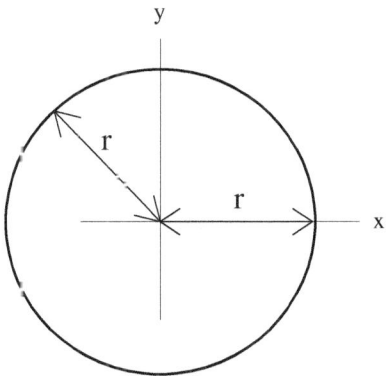

Figure 2.7:

the point $(0,0)$ is to say

$$x^2 + y^2 = r^2 \tag{2.6}$$

The points whose coordinates satisfy this equation lie on the circle, so this is the equation of the circle. Now suppose we look for all events in the xt-plane of Minkowski spacetime which are at a fixed spacelike distance d from the origin (0,0). In other words, we look for all events (x,t) which satisfy the condition

$$x^2 - t^2 = d^2 \tag{2.7}$$

Instead of a circle, we get the curves shown in figure 2.8. Notice also that

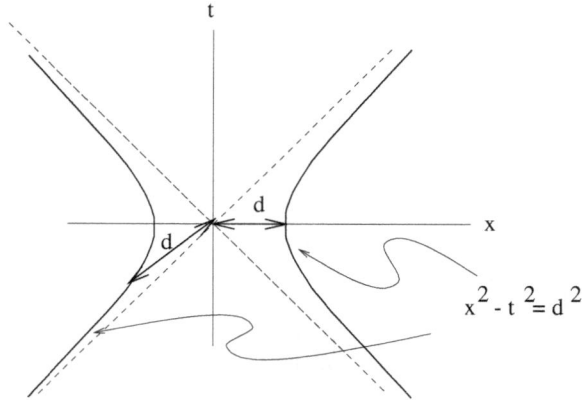

Figure 2.8:

the events (x,t) which are *timelike*-separated from (0,0) by a fixed amount s must satisfy the equation

$$-x^2 + t^2 = s^2 \tag{2.8}$$

They lie on the curves shown in figure 2.9.

The notion of a *null line* plays a leading role in relativity. A line of events is a null line if any two events on the line are null-separated. In the xt-plane the null lines are those lines which make a 45° angle with the coordinate axes. You can understand why that is so by considering any pair of events on such a line, like events E and F in the figure 2.10. In moving from event E to event F, the t-component changes exactly as much as the x-component, so $|t_2 - t_1| = |x_2 - x_1|$ and $\mathcal{I}_{EF} = 0$. The diagram shows a number of null lines. *Any* 45° line in the xt-plane is a null line.

Minkowski Spacetime 21

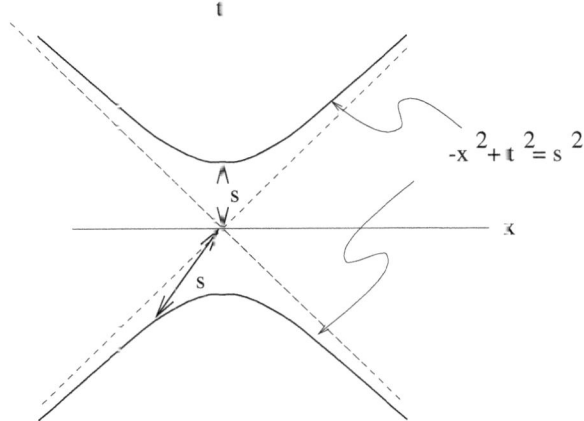

Figure 2.9:

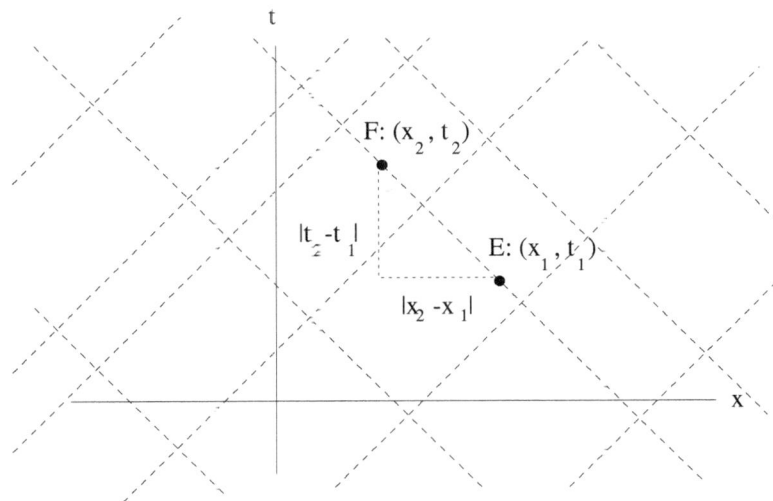

Figure 2.10:

Since the geometry of Minkowski spacetime is not perceived automatically by us human beings, it is necessary to introduce a couple of rules which allow us to see how Minkowski spacetime matches the real world. The first rule links light propagation to the null lines. Electromagnetic radiation, of which visible light is an example, consists of special particles called *photons*. Here is our first rule:

> The worldline of a photon is a null line.

As a photon advances along its null worldline, its x-coordinate changes at the same rate as its t-coordinate. I'll measure time in seconds, so each unit along the t-axis represents one second. Light is known to travel at 186,000 miles per second,[3] so each unit along the x-axis must represent 186,000 miles. That is the distance a photon travels in one second. It is convenient to call this distance a *light-second*. So one light-second is the same as 186,000 miles. I measure timelike separations in seconds and spacelike distances in light-seconds. Using these units, the speed of light is 1, i.e. one light-second per second.

> *Remark*: It is not necessary to use these particular units. Suppose, for example, that you wish to measure timelike separations in seconds and spacelike distances in miles. The Minkowski interval formula must then be changed to
>
> $$\mathcal{I}_{EF} = -(x_2 - x_1)^2 + c^2(t_2 - t_1)^2 \qquad (2.9)$$
>
> where $c = 186,000$ is the speed of light in these units. You can check that two events are null separated ($\mathcal{I}_{EF} = 0$) provided $|x_2 - x_1| = c|t_2 - t_1|$. The worldline of a photon is still a null line, any two events on it being null separated. (With these units, the the x-component changes 186,000 times faster than the t-component along a photon's null worldline. A null line is not a 45° line with this system of units.)
> It's the null lines – possible photon worldlines – which are important in relativity. I have chosen units of seconds and light-seconds so that the null lines are the convenient 45° lines in my diagrams.

---

[3]A more accurate value for the speed of light is 186,290 miles per second.

Minkowski Spacetime

Any straight line in Minkowski spacetime can be classified as a null line, a timelike line, or a spacelike line. As I have said, any two events on a null line are null separated. A *timelike line* is characterized by the property that any two events on it are timelike separated. Sliding along such a line, the t-coordinate changes faster than the x-coordinate (figure 2.11). On a

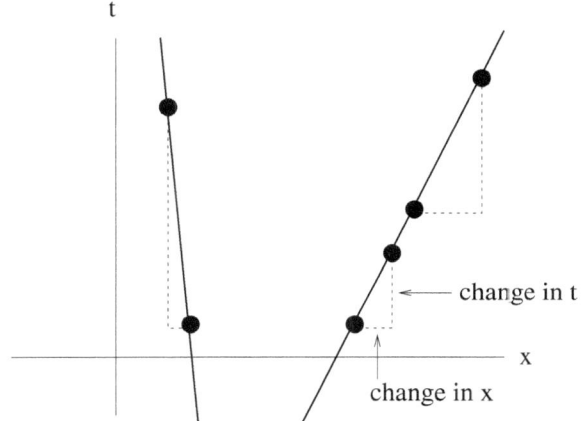

Figure 2.11: Examples of timelike lines.

spacelike line any two events are spacelike separated. Along a spacelike line, the x-coordinate changes faster than the t-coordinate (figure 2.12).

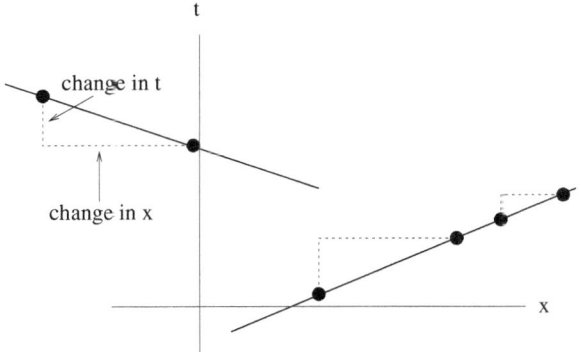

Figure 2.12: Examples of spacelike lines.

It is apparently true that no particle can move faster than light. Searches for faster particles have all met with negative results. This means that, as a particle advances along its worldline, the x-coordinate cannot change faster than the t-coordinate. The worldline of a particle, therefore, cannot be a spacelike line. Ordinary objects have timelike worldlines, and photons have null worldlines.

The separation between any pair of timelike-separated events can, in principle, be measured using a clock. It is only necessary to arrange that the worldline of the clock includes the straight line segment which joins the two events. The timespan recorded by the clock as it passes from one event to the other is equal to the separation between the events. This is another rule which is used in order to understand Minkowski spacetime as the geometry of the real world.

> If the worldline of a clock is a straight line, then the timespan which the clock records between any pair of events on its worldline is equal to the Minkowski separation of the two events.

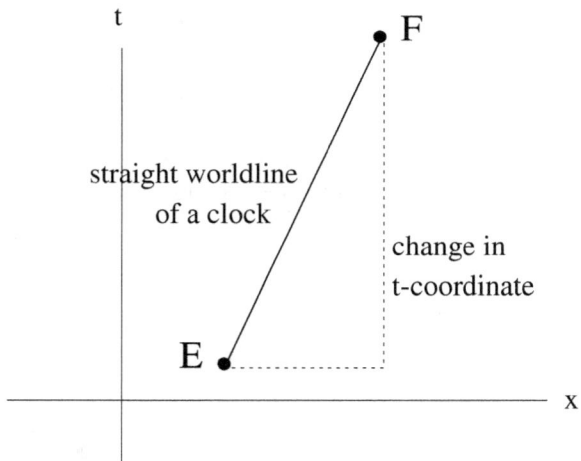

Figure 2.13: $\overline{EF}$ is the time measured by the clock.

In figure 2.13, the separation $\overline{EF}$ is the time measured by the clock. It is *not* equal to the difference in the t-coordinates of events E and F. The separation $\overline{EF}$ is sometimes called the *proper time* of the line segment from E to F in order to distinguish it from the coordinate time (change in t-coordinate).

Due to the fact that particles are limited to speeds less than the speed

Minkowski Spacetime

of light, it is not possible to have instantaneous communication. Signals can propagate no faster than light. Suppose you do something dramatic at event E. The effects of your action can be felt only in the shaded region shown in figure 2.14. What you do at event E cannot affect events which are spacelike-separated from E, nor can it affect events which are timelike-

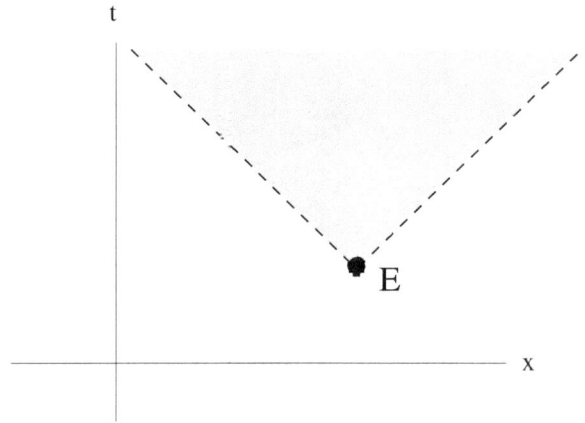

Figure 2.14:

separated from E but in the past. For example, if you set off a flash bulb at event E, photons propagate to the left and to the right at the speed of light. Their worldlines are the dashed lines in figure 2.14, and those lines constitute a boundary for the region of spacetime to which you can send signals from event E.

The speed of light also places limitations on the scope of your knowledge at event E. Events which are spacelike-separated from E cannot send signals to event E, so you cannot have knowledge of what happens at such events if you are at event E. The shaded region in figure 2.15 is the region from which you can receive information at event E.

In 4-dimensional Minkowski spacetime. Imagine once again that you set off a flash bulb at some event E. Photons go outward in all directions at the speed of light. A moment after the flash, the photons are located on a sphere in space which is centered on the flash bulb. As time goes on, the sphere of light gets larger and larger (its radius increasing at the speed of light). Figure 2.16 shows the sphere of photons at two different times. (Each sphere looks like a circle in my picture because I can only represent two spatial dimensions in the sketch.) Each photon has a null worldline. Taken together, these null lines form the *null cone* (or *light cone*) of event

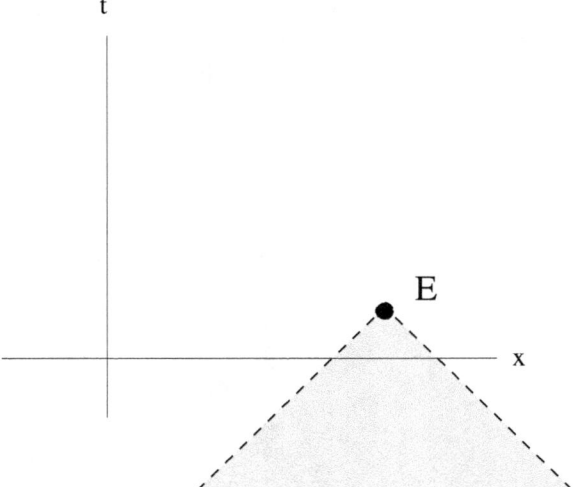

Figure 2.15:

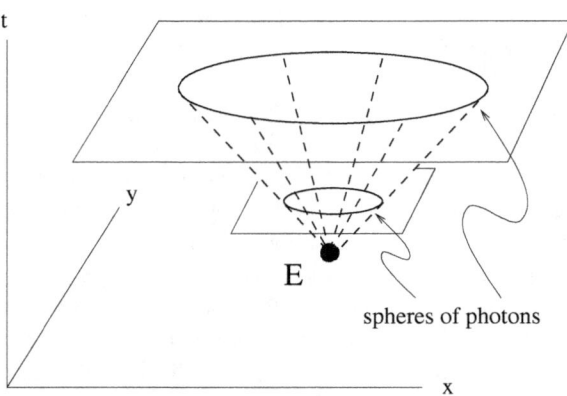

Figure 2.16:

Minkowski Spacetime

E. The interior of the cone consists of events which are timelike separated from E. The interior region in the figure is the part of spacetime to which you can send signals from event E.

The above figure shows only the future null cone of E. Extending the null lines into the past gives the past null cone (figure 2.17). The past null

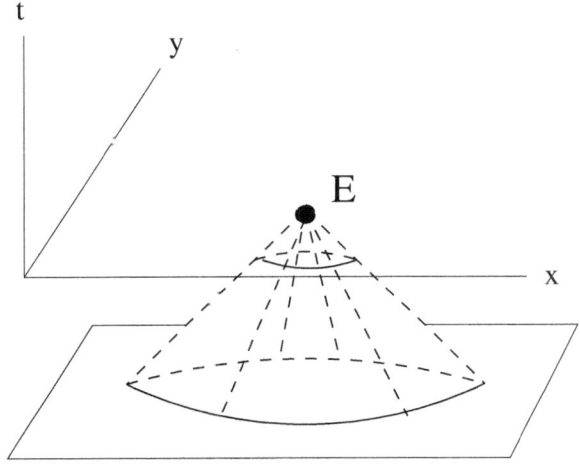

Figure 2.17:

cone together with its interior is the portion of spacetime about which you can have knowledge if you are at event E.

Chapter 3

THE RELATIVITY OF VELOCITY

The name "relativity" arises from the basic fact that one can talk about the velocity (or speed) of an object only *relative* to something else. If I say "I am standing still" or "I am moving 55 miles per hour," you understand that I am talking about my speed relative to the ground beneath me. Even when I am standing still relative to the ground, however, I am moving relative to the sun with some speed, and I'm moving relative to the center of our Milky Way galaxy with some other speed. And our entire galaxy is moving relative to other galaxies.

The essential thing to notice is that, when you say how fast you are moving, you are always reporting your speed *relative to something else*. There is no way to determine your speed without observing some other object for reference, and so your speed is a relative quantity. In contrast, acceleration is *not* just a relative quantity. It is something you can determine without reference to anything else. Whenever your car accelerates, you can feel your seat pushing you forward. The greater your acceleration, the harder your seat pushes you. If you put a bathroom scale behind your back, you can measure the force and deduce how much you are accelerating. Acceleration is meaningful not only relative to something else (such as the ground); it is an absolute quantity. If you were in a space capsule, far away from all stars, you could still feel whether or not you were accelerating. Velocity, however, is not like that. It would be meaningless to ask you simply how fast you and your space capsule were traveling; only relative velocities can be determined.

Nature is egalitarian when it comes to measuring velocities. *Any* object can be regarded as being at rest and velocities of all other objects measured relative to it. There is no preferred reference object. For us terrestrial beings, it is usually convenient to regard the ground as stationary. But not always. For example, suppose you are playing catch with a frisbee inside a railway car. You and your friend and the air around you are all moving with the train relative to the ground. It is natural, though, to regard yourself as stationary and think of the frisbee's motion relative to yourself rather than thinking about its speed relative to the ground. A less homely example is that of the Apollo astronauts landing on the moon. As they settled down to the moon's surface they were moving more than 2,000 miles per hour relative to the ground on Earth. It was natural for them to regard the moon's surface as stationary and measure their speed relative to it.

Because of the relativity of velocity, any non-accelerating person can regard himself as stationary and measure the velocities of all objects relative to himself. Minkowski spacetime manifests this egalitarianism: the geometry appears the same to all non-accelerating persons. In other words, the Minkowski geometry does not discriminate between non-accelerating persons; there is no special reference person. I'll be more specific about this in the chapter on "Switching Reference Systems."

Chapter 4

RADAR MEASUREMENTS AND SIMULTANEITY

We use clocks to measure timelike separations, but how do we measure spacelike distances? If you want to know how far it is to the moon, you can bounce a radar signal off the moon and see how long it takes for it to return. (Radar, like visible light, consists of photons. A radar photon is not energetic enough to be detected by your eye, but, being electromagnetic radiation, it moves at the speed of light.) The radar signal reflected from the moon arrives back after 2.6 seconds. So it takes 1.3 seconds to get to the moon and 1.3 seconds to return. That tells you that the moon is 1.3 light-seconds away. (Notice that the light-second is a natural distance unit when radar is used to make measurements.)

Radar is an accurate way to measure distances, even ordinary distances on Earth. This is because there exist highly accurate clocks which can measure very precisely the timespan it takes for the radar photons to go and return from some object. Measuring distances this way is far more accurate than laying down yard sticks end to end.

Let's make a spacetime diagram which shows a radar measurement. We can choose to think of you as stationary and sitting at the origin of spatial coordinates. In other words, your worldline will be the t-axis (figure 4.1. Now suppose there is an object at some distance from you which is not moving relative to you. Its worldline is then parallel to yours. If you bounce a radar signal off the object, the spacetime diagram will look like figure 4.2

Radar Measurements and Simultaneity

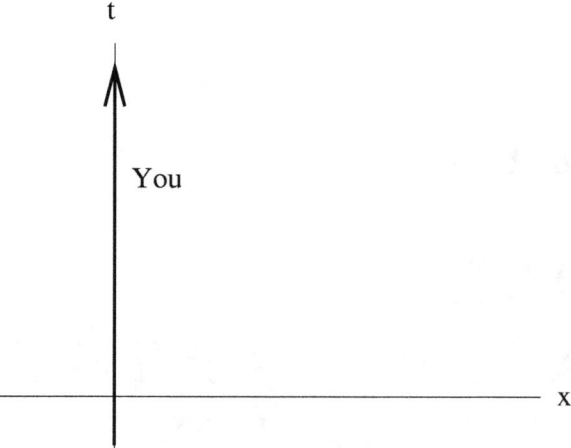

Figure 4.1:

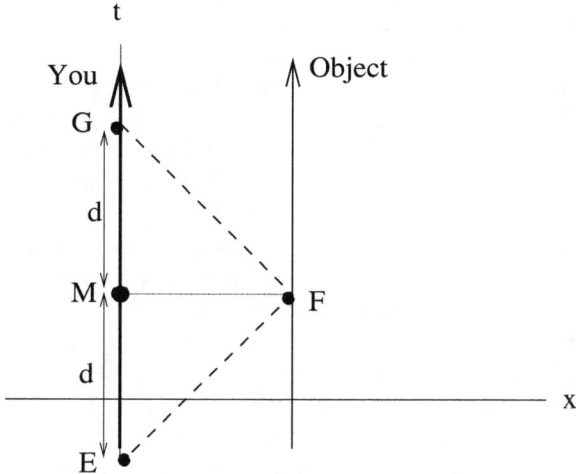

Figure 4.2:

Radar Measurements and Simultaneity   33

The photons are sent out at event E, bounce off the object at event F, and return to you at event G. Suppose your clock measures the separation $\overline{EG}$ to be some number of seconds which I'll call $2d$. You would deduce that the photons took d seconds to get to the object and d seconds to return. So the object is at a distance of d light-seconds from you. Let M be the event halfway between events E and G on your worldline. By measuring the separation $\overline{EG}$ to be $2d$ seconds, you have determined that the distance $\overline{MF}$ is d light-seconds.

> *Recreation*: It is important to verify that this radar measurement does indeed give the correct Minkowski distance $\overline{MF}$. (1) Using the methods of chapter 2, check that, for two events M and G on the t-axis, the separation $\overline{MG}$ is simply the absolute value of their t-coordinate difference: $\overline{MG} = |t_G - t_M|$. (Here t_G stands for the t-coordinate of event G and t_M is the t-coordinate of M.) (2) Check that, for two events M and F which have the same t-coordinate, the distance $\overline{MF}$ is simply the absolute value of their x-coordinate difference: $\overline{MF} = |x_F - x_M|$. (3) Recall that, for two events on a photon's (null) worldline, the x-coordinates differ by the same amount as the t-coordinates in absolute value. So, for events F and G in figure 4-2, $|x_G - x_F| = |t_G - t_F|$. Now use the above results, together with $t_F = t_M$ and $x_G = x_M$, to check that $\overline{MF} = \overline{MG}$. Therefore, if the separation $\overline{MG}$ is measured to be d seconds, then the Minkowski distance $\overline{MF}$ must indeed be d light-seconds.

In the spacetime diagrams above, I have drawn the worldlines of you and the object as straight lines – no wiggles or kinks. Any straight line in the xt-plane has the following property: Sliding along the line, the x-coordinate changes by the same amount v during *any* one unit of coordinate time. See figure 4.3. The worldline of a non-accelerating object should have precisely that property. Such an object has a constant speed v relative to a person whose worldline is the t-axis.

Now suppose you are using radar to measure the distance to some object, say a rocket, which is moving relative to you. You may still regard yourself as stationary and draw your worldline as the t-axis. If the rocket is moving away from you with speed v, then it moves an amount v away from you during each unit of time. A single radar measurement then looks like figure 4.4. From this measurement you would deduce the rocket's distance to be $\frac{1}{2}\overline{AC}$ when you are at event O. That event occurs halfway between the time when you emit the radar signal and the time when you receive the

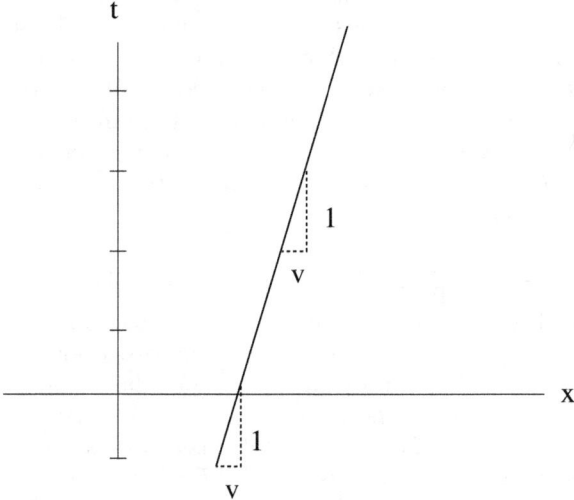

Figure 4.3:

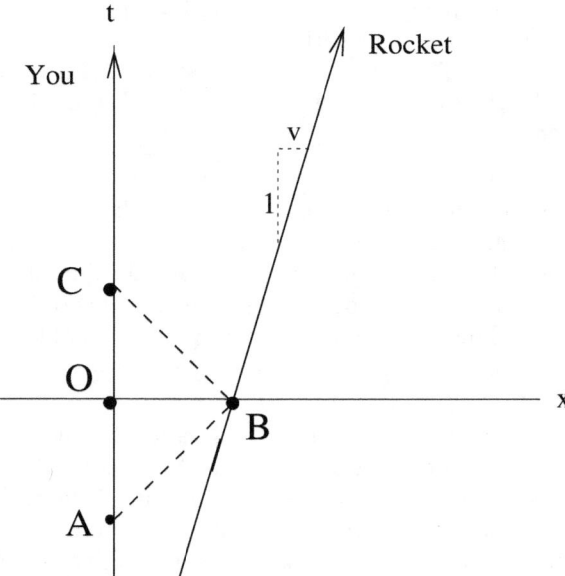

Figure 4.4:

return signal.

The distance from you to the rocket changes, of course, because the rocket is moving relative to you. A moment later, when you are at event M and the rocket is at event F, you would find the distance to the rocket to be $\frac{1}{2}\overline{EG}$ by performing the measurement shown if figure 4.5.

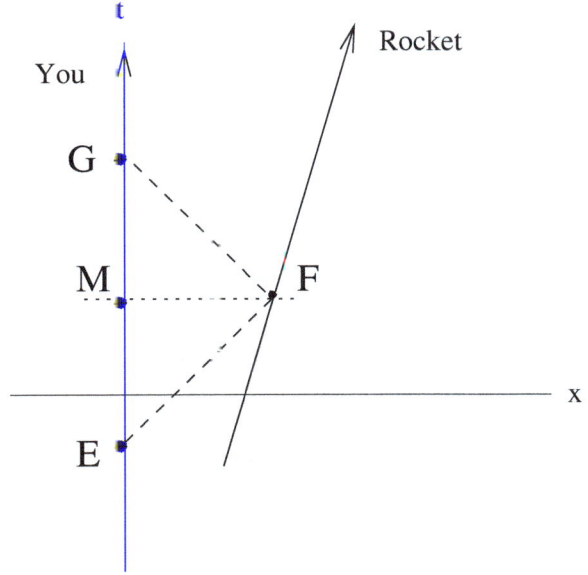

Figure 4.5:

I'd like you to notice a consequence of our rule which says that any photon has a null worldline. When the photon bounces off a rocket, it follows the 45° line away from that event. It doesn't matter whether the rocket is moving toward you at high speed, moving away from you, or stationary relative to you. The speed of the photon is 186,000 miles per second before and after it bounces off the rocket. See figure 4.6. In this sense, a photon is quite different from a baseball, for example. When a baseball bounces off a bat, its speed depends very much on how fast the bat is moving (relative to the pitcher, say). A photon behaves in this respect more like a sound wave in air. A sound wave travels 770 miles per hour relative to the air, even if it has just bounced off an object which is moving through the air. Please note, in particular, that a radar photon always takes exactly as long to travel to an object as it takes to return to you from the object. (If its speed were different after bouncing off the object, then this would not be the case.) From this fact you can deduce that *the*

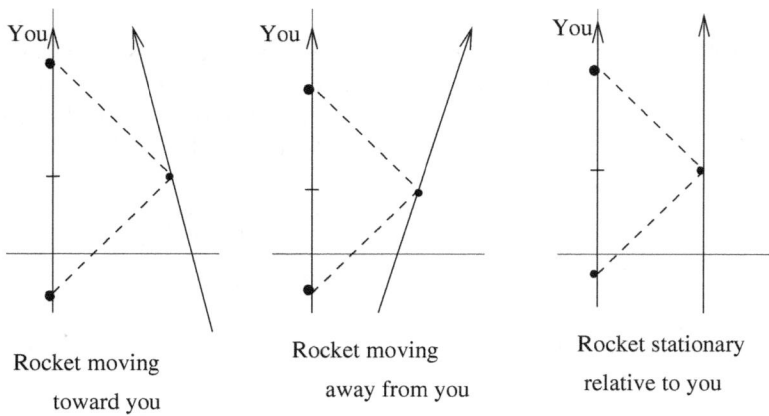

Figure 4.6:

reflection of the photon at the object occurs simultaneously with the event on your worldline which is halfway between the emission of the photon and its return.

In the radar measurements sketched in figures 4.4 and 4.5, you regard event O as simultaneous with event B (at which the first radar signal is reflected) and event M as simultaneous with event F (at which the second radar signal is reflected). *The rocket pilot disagrees*!! From her point of view, event O occurs after event B and event M occurs after event F. There are many ramifications of this fact that two persons in relative motion disagree about which events are simultaneous. In later chapters we'll see that it leads directly to the effects known as "time dilation" and "Lorentz contraction." It is therefore worth our while to examine this matter of simultaneity in some detail.

The rocket pilot may regard herself as stationary, with you moving with speed v relative to her. You may think she is being unrealistic about this if you are sitting still on the Earth and she is flying along in a small rocket. The notions of the previous chapter, however, must be taken seriously. Anybody can be chosen as a stationary reference, and the speeds of all other objects measured relative to that person. It may help to imagine that the rocket pilot is motionless relative to the sun. After all, we are accustomed to thinking of the sun as stationary, with the Earth moving relative to it. So let's imagine that you (and the Earth) are moving with speed v away from a rocket which is stationary (relative to the sun).

Suppose the rocket pilot makes a radar measurement by bouncing a photon off your belt buckle. Her clock measures a certain timespan for the round trip of the photon. From her point of view, the photon takes just as long to get from the stationary rocket to you as it takes to return to the rocket. She deduces that the photon's reflection occurs halfway through the measured timespan. In other words, the rocket pilot regards the reflection as simultaneous with the event on her worldline which is halfway between the emission and the return of the radar photon.

Let's look at a spacetime diagram of a radar measurement she might make. The radar bounces off of you at event O in figure 4.7. It is event N

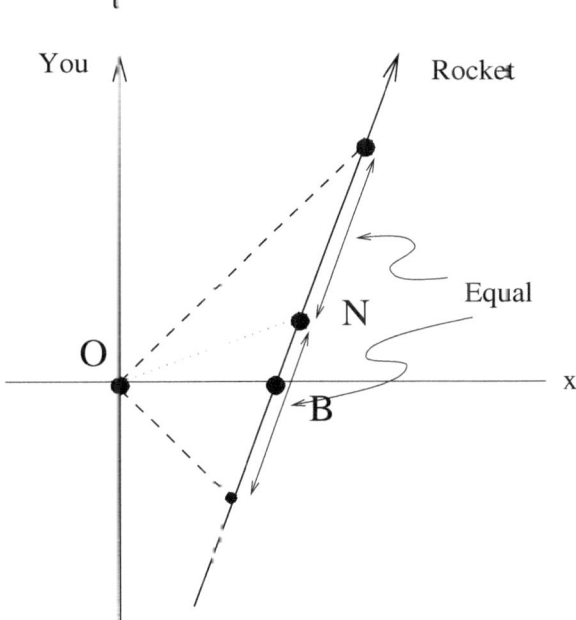

Figure 4.7:

on her worldline which occurs halfway between the emission of the signal and the reception of its reflection. So she regards the reflection at O as occurring simultaneously with event N, somewhat after the occurrence of event B. You regard event B as simultaneous with event O, whereas she regards event N as simultaneous with O.

The event on your worldline which she regards as simultaneous with event B is some event K, as shown in figure 4.8. In this measurement, B *does* occur midway between emission of the radar and its return, but the signal bounces off of you at event K, prior to event O.

Similarly, as exhibited in the diagrams of figure 4.9, she regards M as

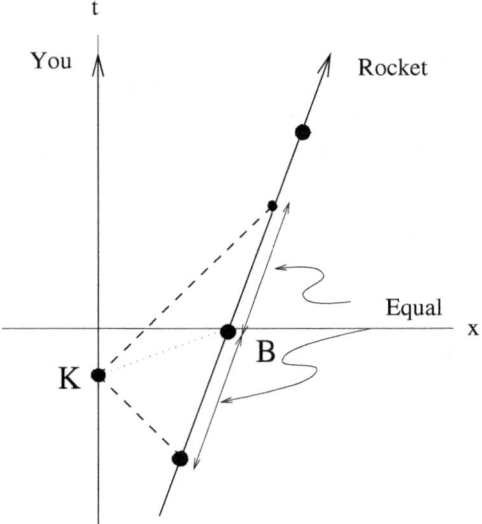

Figure 4.8:

simultaneous with some event P (instead of F), event F being regarded as simultaneous with an event L, which is plotted in the second diagram.

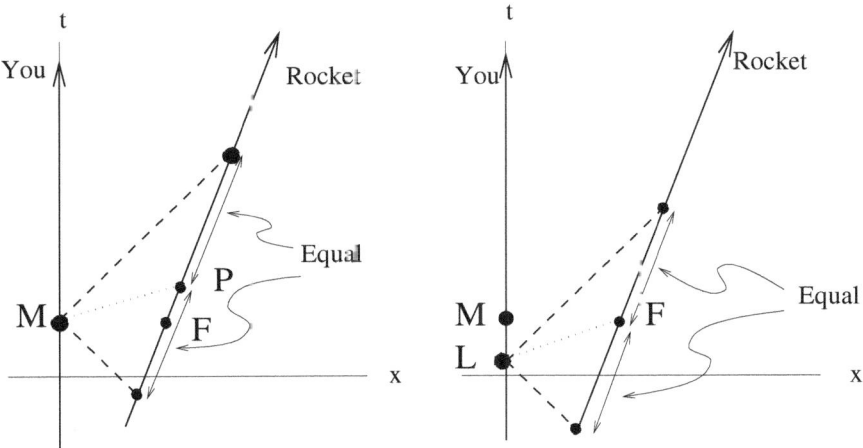

Figure 4.9:

Remark: There is a more direct way to see that persons in relative motion do not agree about simultaneity. We can deduce it immediately from the combination of just two facts: (1) the rule that clocks measure Minkowski timelike separations and (2) the fact that two persons agree as to their relative speed.

Suppose a rocket passes you at event O, traveling with relative speed v. (See figure 4.10.) Let K be the event on your worldline which occurs 1 second after event O, and let N be the event on the rocket's worldline which you regard as simultaneous with K. During 1 second on your clock, the rocket moves v light-seconds away from you. That's what it means to say that the rocket's speed relative to you is v. The coordinates for event N are therefore $(v, 1)$. In advancing from event O to event N, the rocket's clock records the separation $\overline{ON}$, and you can check that this is equal to $\sqrt{1-v^2}$. When the rocket pilot is at event N, she would say that you are at a distance of $v\sqrt{1-v^2}$ light-seconds, because you have been traveling at speed v for $\sqrt{1-v^2}$ seconds. But the distance $\overline{KN}$ is equal to v, not $v\sqrt{1-v^2}$. We must conclude that, when the rocket is at event N, the pilot does not think you are at event K. Instead, she would say you are at event F whose coordinates are $(0, 1-v^2)$. (See figure 4.11.) You can check that $\overline{FN}$ is indeed equal to $v\sqrt{1-v^2}$. According to the rocket pilot, therefore, it is event F which is simultaneous with event N.

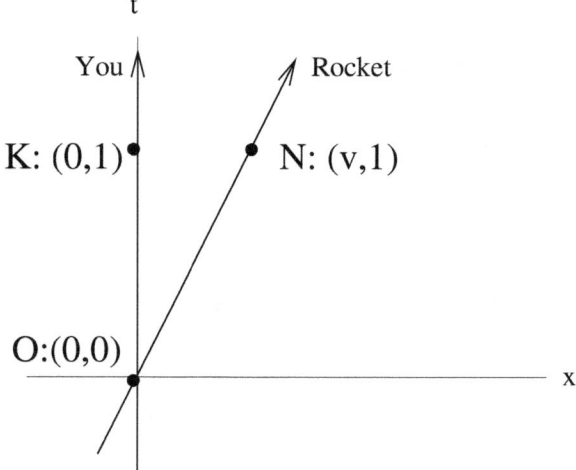

Figure 4.10:

> *Recreation*: Please refer to figure 4.11. After $1 - v$ seconds on your clock, the rocket has gone a distance $v(1 - v)$ relative to you. Let E be that event with coordinates $(v - v^2, 1 - v)$. Let G be the event on the rocket's worldline with coordinates $(v + v^2, 1 + v)$. Show that $\overline{EF} = 0$ and $\overline{FG} = 0$, so a radar measurement by the rocket pilot might look like figure 4.12. In the preceding "remark," events F and N were found to be simultaneous. Check that this is consistent with our original criterion for simultaneity by showing that N is halfway between E and G. In other words, verify that $\overline{EN} = \overline{NG}$.

Let's now try to find all the events in the xt-plane which you regard as simultaneous with a particular event M on your worldline if your worldline is the t-axis. Any such event F is distinguished by the possibility of a radar signal which originates at some event E on your worldline, is reflected at F, and meets your worldline again at an event G, with $\overline{EM} = \overline{MG}$ as in figure 4.5. Now let F be any old event. We can easily check whether or not you regard F as simultaneous with M. We simply draw the two null lines in the xt-plane which pass through the event F. The null lines meet your worldline in two events; call them E and G. You regard the event F as simultaneous with M provided $\overline{EM} = \overline{MG}$. (See figure 4.13.)

You can readily see that all events which you regard as simultaneous

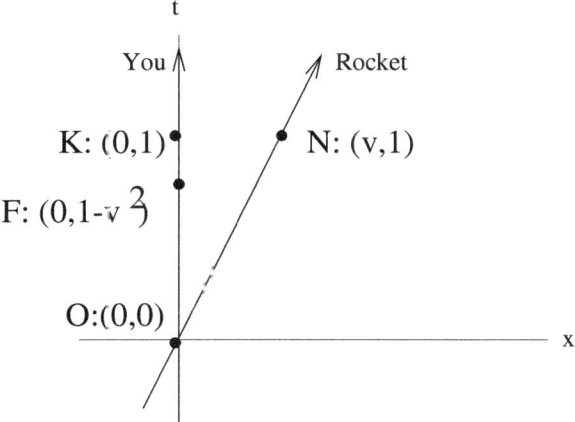

Figure 4.11:

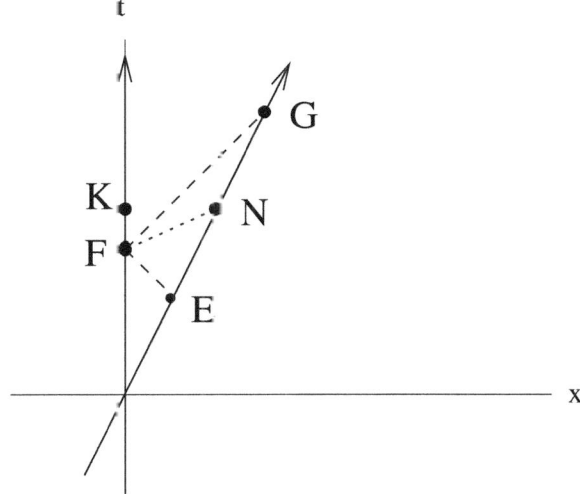

Figure 4.12:

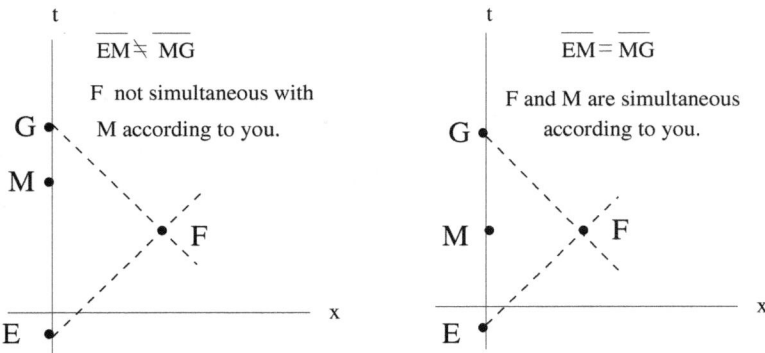

Figure 4.13:

with M form a line through M on which the coordinate t is constant. This is a very agreeable answer. The events which you regard as simultaneous

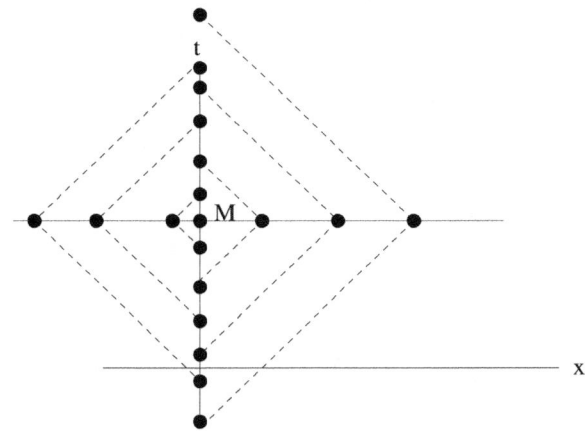

Figure 4.14:

with M are those events which all occur at the same coordinate time t. (Your worldline is the t-axis. See figure 4.14.)

Now consider a similar question: What events does the rocket pilot regard as simultaneous with some event N on her worldline? We can test any event F for *this* property. Draw the two null lines through F. The two null lines meet the rocket's worldline in two events which we can again call E and G. The rocket pilot regards events F and N as simultaneous provided $\overline{EN} = \overline{NG}$. Figure 4.15 is the diagram for testing an event F.

Radar Measurements and Simultaneity

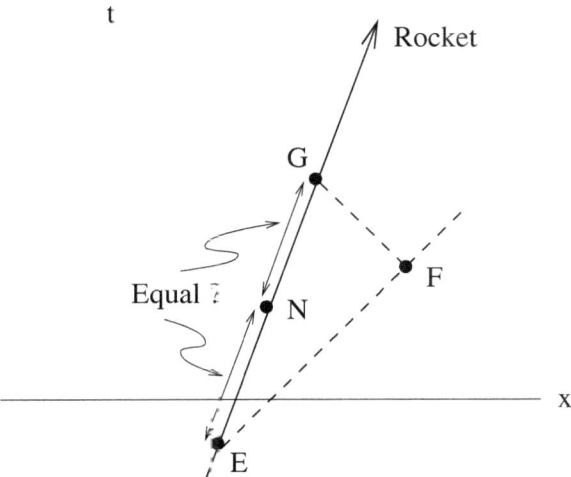

Figure 4.15:

The events which do satisfy this criterion form a line, but it is not a line on which t is constant. See figure 4.16. This new line is a line of simultaneous events according to the rocket pilot.

Pilots of rockets at different speeds have different notions of simultaneity. The sketches in figure 4.17 show the worldlines of rockets with various speeds relative to you. Each diagram shows also the line of events which the pilot would regard as simultaneous with event N. The faster the rocket is moving relative to you, the more the line of simultaneity is tilted with respect to a (horizontal) line of constant t. In each picture of the previous figure, the dotted line (a 45° line) bisects the angle between the worldline and the line of simultaneity. So, if you are handed a rocket's worldline with some event N on it and someone asks you to find the line of events which the pilot regards as simultaneous with N, you can answer pictorially quite easily. You just draw the 45° dotted line and then find the line on the other side which is symmetrical to the worldline, as illustrated in figure 4.18.

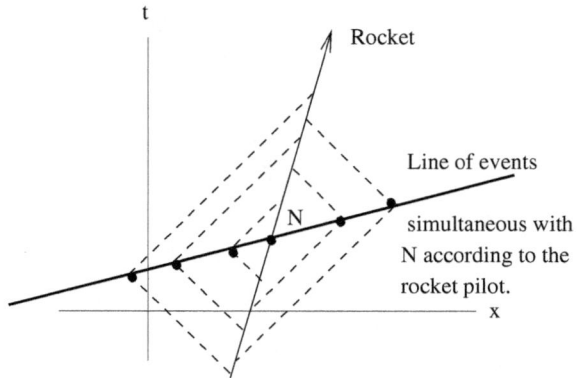

Figure 4.16:

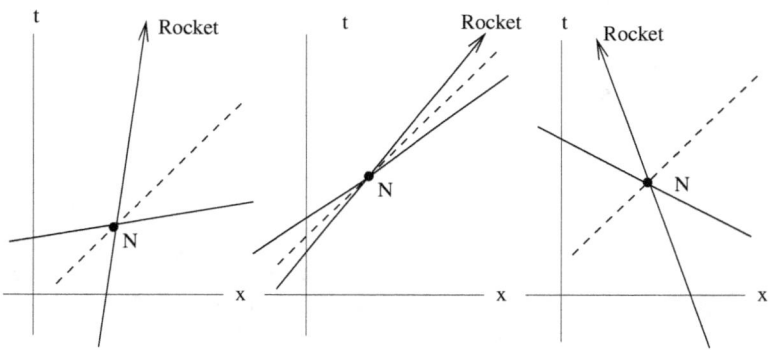

Figure 4.17:

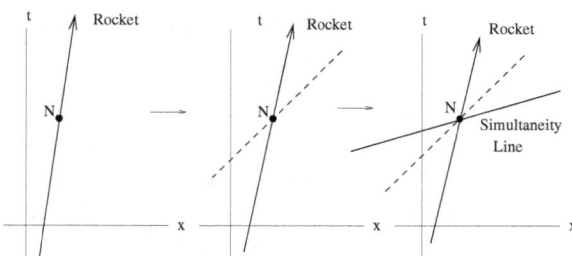

Figure 4.18:

Radar Measurements and Simultaneity

> *Explanation:* If you recall that the null lines in the xt-plane are at $45°$ to the coordinate axes, then you see that null lines in the plane always form right angles where they meet each other. (Note: When I speak of null lines making $45°$ angles with the coordinate axes or two null lines meeting in a right angle, I am referring to angles which you would measure in the ordinary way using a protractor on the sketches.) The radar construction thus produces rectangles, with the rocket worldline and line of simultaneity as diagonals. See figure 4.19. The (dotted) line which bisects the angle between the diagonals is parallel to the sides of the rectangles. Since these sides are null lines at $45°$ to the coordinate axes, the dotted bisecting line must also be a $45°$ line.

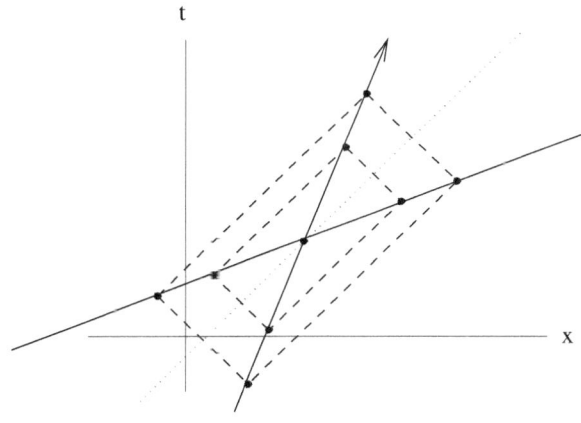

Figure 4.19:

Finding the line of simultaneity through a particular event is a useful technique, and we should figure out how to do it algebraically. As a first step, notice that a line in the xt-plane can be characterized by an equation of the form

$$t = mx + b \tag{4.1}$$

In other words, an event with coordinates (x, t) is on the line if and only if its coordinates x and t satisfy $t = mx + b$, where m and b are fixed numbers.

Example: If $b = 2$ and $m = \frac{2}{3}$, then the events (x,t) satisfying $t = \frac{2}{3}x + 2$ form the line in figure 4.20. (Three particular events on the line are labeled.)

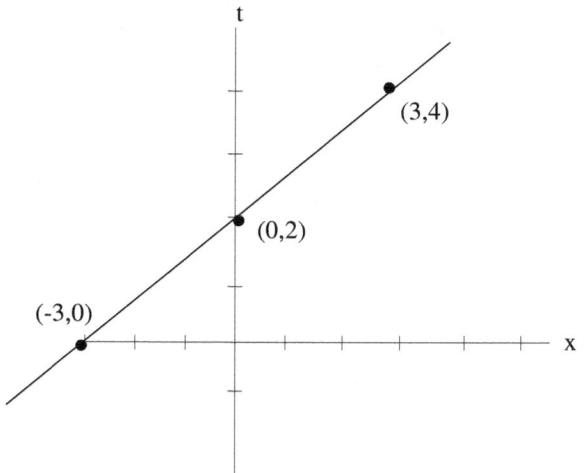

Figure 4.20:

The number b is the value of t where the line meets the t-axis. This is because $x = 0$ at the t-axis, so $t = mx + b$ is there the same as $t = b$. This number b is called the *t-intercept*. The number m is called the *slope* of the line. Sliding along the line, m is the amount the t-coordinate changes while the x-coordinate increases by 1 unit. See figure 4.21. The line's slope and where the line crosses the t-axis are all you need to know in order to draw the line. (Vertical lines in the xt-plane are exceptions. The equation for such a line is simply $x = a$ where a is a fixed number.)

Once again, let me draw the worldline of a rocket moving relative to you with speed v, and I'll mark an event N on the worldline. See figure 4.22. During any one unit of coordinate time t, the rocket's x-coordinate changes by an amount v. This means that t changes by an amount $\frac{1}{v}$ as x increases by 1 unit. That tells us that the slope of the rocket's worldline is $\frac{1}{v}$. Now let's draw in a dotted 45° line through N, and find the symmetrical simultaneity line (figure 4.23).

The figure shows that, in sliding along the simultaneity line, the t-coordinate changes by an amount v as the x-coordinate increases by 1 unit. The slope of the simultaneity line is therefore equal to v, the speed of the rocket

Radar Measurements and Simultaneity

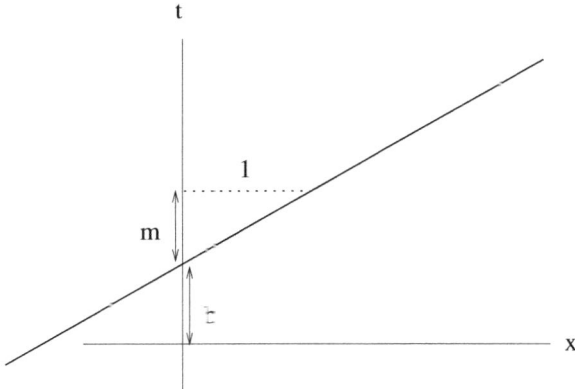

Figure 4.21:

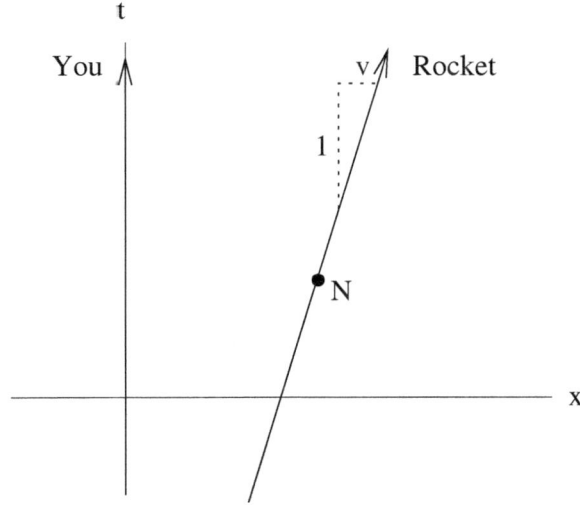

Figure 4.22:

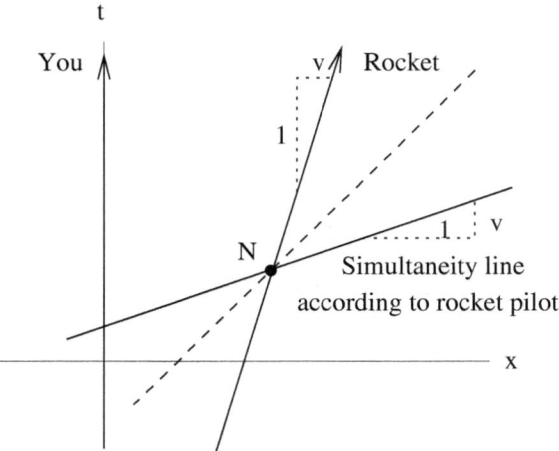

Figure 4.23:

relative to you.

Let me repeat this important result. If a rocket has a worldline with slope $\frac{1}{v}$, the pilot regards a line of events as simultaneous provided the line has slope v.

There are lots of such lines; in fact one such line passes through any event of the rocket's worldline. See figure 4.24. Suppose, however, we want the *one* line of events which are all simultaneous with event N. How do we find the equation for that particular line? Well, suppose N has coordinates (3,2) for example. Because the line in question has slope v, its equation is $t = vx + b$, for some number b. Since N itself is on this line, (3,2) must satisfy the equation, so $3 = v \cdot 2 + b$. Solving this for b gives $b = 3 - 2v$, so the equation of the line is

$$t = vx + (3 - 2v) \qquad (4.2)$$

where v is the rocket's speed relative to you.

> *Recreation*: Suppose a rocket is moving at two-thirds the speed of light relative to you, so $\frac{1}{v} = \frac{3}{2}$. Suppose you and the rocket were together at the event (0,0). Notice that the worldline of the rocket has equation $t = \frac{3}{2}x$, and the event (6,9) is on the worldline. Find the equation for the line of events which the pilot regards as simultaneous with (9,6). Answer: $t = \frac{2}{3}x + 5$).

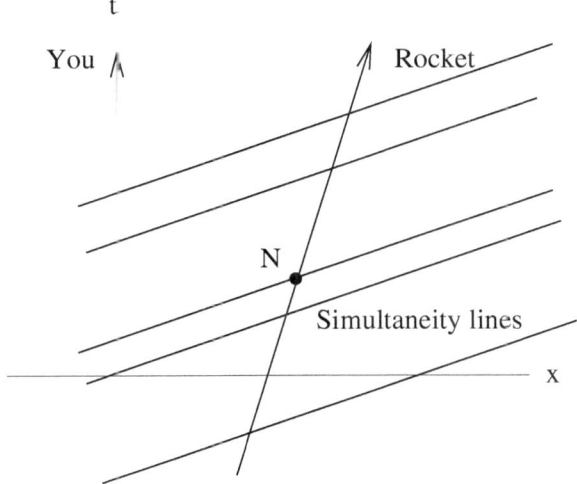

Figure 4.24:

Recapitulation: Because of the relativity of velocity, any person is justified in thinking of himself as motionless and measuring velocities of all objects relative to himself. To measure the distance to an object, the person bounces a photon off it and measures the timespan during which the photon makes its round trip. The distance in light-seconds is one-half the timespan measured in seconds. Since the photon takes just as long to get to the object as it takes to return, the person regards the mid-point of the timespan as the instant when the photon is reflected. In other words, he says the event at which the photon is reflected is simultaneous with the event on his worldline which is midway between the sending of the photon and its return. His notion of simultaneity is not shared by persons moving relative to him. If the worldline of a person has slope $\frac{1}{v}$ in the xt-plane, that person regards any line with slope v as a line of simultaneous events.

We should check to make sure that distances as measured by radar are the same as the Minkowski distances defined in chapter 2. We already verified this for measurements made by a person (you) whose worldline is the t-axis. (Recall the "recreation" following figure 4-2.) Let's check that it is true also for an arbitrary rocket pilot.

Once again, please examine the spacetime diagram of a radar measurement by a rocket pilot (figure 4.25). As usual, the photon starts at event E, is reflected at F, and arrives back at the rocket at G. Event N is midway between E and G, so $\overline{NG} = \frac{1}{2}\overline{EG}$. According to the way radar measure-

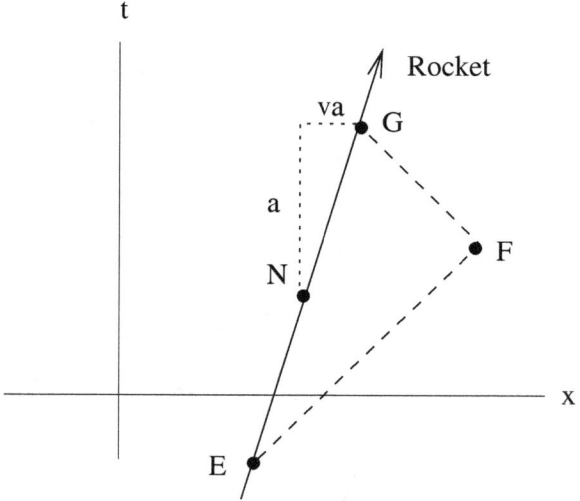

Figure 4.25:

ments are interpreted, the rocket pilot concludes that the distance $\overline{NF}$ is equal to $\overline{NG}$, which is one-half the measured timespan $\overline{EG}$. We wish to check that this conclusion is correct. That is to say, if we use the Minkowski interval formula to compute $\overline{NF}$ and $\overline{NG}$ as in chapter 2, we should find that they are equal. The next "calculation" is one way to do this check on the validity of radar measurements. The "recreation" following is an alternative method.

Calculation: In figure 4.25 I have denoted by a the difference in the t-coordinates of events N and G. Since the rocket moves with speed v relative to you, the x-coordinates of events N and G must differ by the amount va. These differences can be used to compute $\overline{NG}$ by the methods of chapter 2:

$$\overline{NG} = \sqrt{-(x_G - x_N)^2 + (t_G - t_N)^2} = \sqrt{-v^2a^2 + a^2} = a\sqrt{1-v^2}$$

To compute $\overline{NF}$, we first symmetrize parts of the figure about a 45° dotted line to obtain figure 4.26. Notice that the t-coordinates of N and F differ by va, while the x-coordinates differ by a. Therefore, the spacelike distance $\overline{NF}$ is

$$\overline{NF} = \sqrt{(x_F - x_N)^2 - (t_F - t_N)^2} = \sqrt{a^2 - v^2a^2} = a\sqrt{1-v^2}$$

It is indeed true that $\overline{NF} = \overline{NG}$.

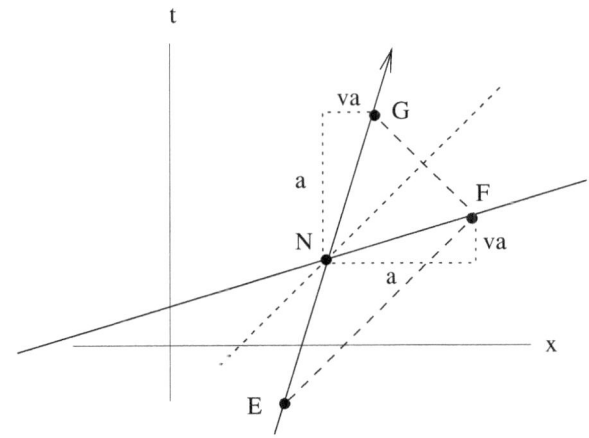

Figure 4.26:

Recreation: Let the coordinates of N be (p, q).

1. Verify that the rocket's worldline has the equation $t = \frac{1}{v}x + (q - \frac{1}{v}p)$.

2. Verify that the event $(p + va, q + a)$ satisfies this equation, so these can be the coordinates for an event G on the rocket's worldline.

3. Use the coordinates for N and G to compute $\overline{NG}$.

4. Show that the line of simultaneity[1] NF has equation $t = vx + (q - vp)$.

5. The line FG has slope -1 and contains event G. Show that its equation is $t = -x + (q + p + a + va)$.

6. Event F is the one event which satisfies the equations for the two lines FG and NF. Solve these two equations together to find the coordinates of event F. (Answer: $(p + a, q + va)$)

7. Now use the coordinates of F and N to compute $\overline{NF}$, and check that $\overline{NF} = \overline{NG}$.

Chapter 5

THE ONE AND ONLY SPEED OF LIGHT

The speed of light is 186,000 miles per second. Period. I don't have to say its speed is 186,000 miles per second relative to the ground or relative to the sun or relative to my favorite rocket pilot or anything else. *Anybody who measures the speed of a photon will find that it is moving 186,000 miles per second relative to himself!*

To verify that the speed of light is the same relative to all persons, we need only examine our familiar diagram of a radar measurement by the rocket pilot. Think of what happens from the pilot's point of view. Exactly

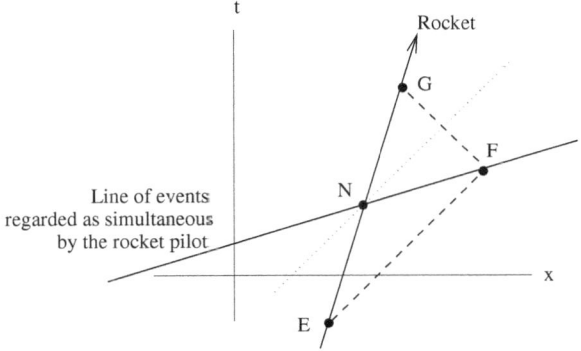

Figure 5.1:

as event N occurs, a photon leaves an object at event F, a distance $\overline{NF}$ from her. She receives the photon at event G, which occurs $\overline{NG}$ seconds after event N according to her clock. From the "calculation" at the end of the previous chapter, we know that $\overline{NF}$ is equal to $\overline{NG}$. Since the distance $\overline{NF}$ traveled by the photon is equal to the measured time $\overline{NG}$ for the photon's flight, she computes the photon's speed to be 1 spacelike distance unit per second. One spacelike distance unit is 186,000 miles, so she measures the photon to travel at a speed of 186,000 miles per second. Recall that you, with your worldline along the t-axis, also determine that the photon moves 1 spacelike distance unit per second. So you and the rocket pilot both measure the photon's speed to be 186,000 miles per second. Since I did not specify any particular value for the speed v of the rocket, it must be true that *any* rocket pilot would measure a photon's speed to be 186,000 miles per second.

This justifies the use of the term light-second to mean 186,000 miles. If different persons were to measure different speeds for photons, then they would not agree on how far a photon travels in one second, and the term light-second would hardly be useful. In the real world, however, everyone finds that a photon travels 186,000 miles per second relative to himself. As a result, any distance measurement can be made by radar and expressed simply in light-seconds. That distance can be converted unambiguously into miles if you so choose.

> *Remark*: We have used the Minkowski spacetime geometry to deduce that all persons measure photons to move at the one and only speed of light. Historically, this fact was discovered in experiments by Michelson and Morley[1] some years before Einstein's first paper on relativity. It was Minkowski spacetime's ability to "explain" the Michelson-Morley result which originally convinced physicists that Minkowski spacetime is real.

It is quite remarkable that everyone measures a photon to move at 186,000 miles per second relative to himself. Imagine please that you are the engineer of a railroad train moving down the track at 70 miles per hour. You wish to alert a person who is working on the track ahead. Suppose you give a blast on your horn. The sound moves through the air at the speed of sound, 770 miles per hour. The worker, stationary relative to the air, would measure the sound as moving 770 miles per hour relative to himself. Since you are moving relative to the air, you would measure the forward-moving sound as going only about 700 (i.e. 770-70) miles per hour relative to you. The speed of the horn signal relative to you is different from its speed relative to the worker.

Next, suppose you fire a flare toward the worker, using a flare gun which hurls the flare at a speed of 200 miles per hour relative to the gun. Since you are holding the gun, the flare moves relative to you at a speed of 200 miles per hour. Since it is fired from the train which is moving toward the worker, the flare's speed is about[2] 270 miles per hour relative to the worker. The speed of the flare relative to you is different from the speed of the flare relative to the worker.

But now suppose you try to alert the worker by flashing the headlight. The photons travel forward from the train at the speed of light – 186,000 miles per second – relative to the train. The worker would measure the photons as moving relative to him at the speed of light – 186,000 miles per second. That is the one and only speed of light. In this respect, a photon differs from an ordinary object (such as a flare), and it differs from a wave in a medium (such as sound). The observer-independence of the speed of light is counter-intuitive to most people, but it is an experimental fact. Minkowski spacetime geometry provides the way to understand it.

> Note: A spacetime diagram of the engineer flashing the headlight at the worker is shown in figure 5.2. In the diagram, $\overline{EO} = \overline{OF}$, so the worker sees the photon move 1 spacelike distance unit per second. Event G is simultaneous with F, according to the engineer, and the separation $\overline{EG}$ is equal to the distance $\overline{GF}$. So the engineer also would say the photon travels 1 spacelike distance unit per second.

---

[2] Adding 70 to 200 is not quite the correct way to calculate the speed of the flare relative to the worker. The next chapter will explain why it is not 100% accurate.

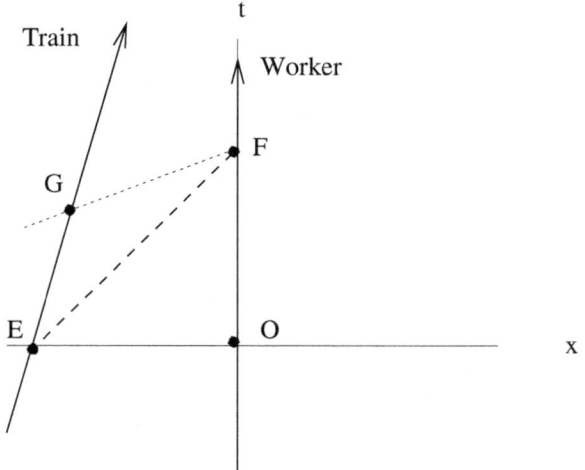

Figure 5.2:

Remark: While figure 5.1 is at hand, there is one other property which I'd like to mention. First of all, note that $\overline{NF}$ is a spacelike distance, so $\mathcal{I}_{NF} = -\overline{NF}^2$, where $\mathcal{I}_{NF}$ is the interval between N and F. We also know that $\mathcal{I}_{NG} = \overline{NG}^2$ and $\mathcal{I}_{FG} = 0$. Now, using the fact that $\overline{NG} = \overline{NF}$, we can deduce this equality:

$$\mathcal{I}_{NF} + \mathcal{I}_{NG} = \mathcal{I}_{FG} \tag{5.1}$$

Compare this equation with the equation of Pythagoras:

$$\mathcal{D}_{MP}^2 + \mathcal{D}_{MQ}^2 = \mathcal{D}_{PQ}^2 \tag{5.2}$$

If this Pythagorean equation is satisfied for a triangle PMQ in Euclidean space, then one says that the line MP is *perpendicular* to the line MQ. (The triangle is called a right triangle. See figure 5.3)

This Pythagorean condition has a generalization in Minkowski spacetime. If the equation

$$\mathcal{I}_{NF} + \mathcal{I}_{NG} = \mathcal{I}_{FG} \tag{5.3}$$

is satisfied for a triangle FNG in Minkowski spacetime, then the lines NG and NF are said to be *orthogonal*. (See figure 5.4.) For lines lying in a spacelike plane (like the xy-plane), orthogonality is the same as perpendicularity. In a timelike plane (like the xt-plane), orthogonality is somewhat different. Two lines in the xt-plane are orthogonal provided they are symmetrical about some null line (a 45° line). In other words, if you have a pair of orthogonal lines in the xt-plane, one of them can be interpreted as a timelike worldline and the other as a line of simultaneity with respect to that worldline.

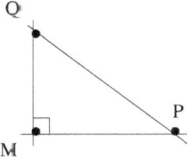

Figure 5.3:

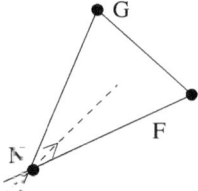

Figure 5.4:

> Note: A null line is orthogonal to itself. Since any two events on the line have a zero interval, any three events form a degenerate triangle (figure 5.5) which satisfies the condition:
>
> $$\mathcal{I}_{NF} + \mathcal{I}_{NG} = \mathcal{I}_{FG} \qquad (5.4)$$

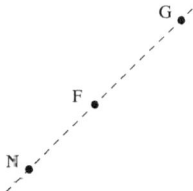

Figure 5.5:

Chapter 6

SWITCHING REFERENCE SYSTEMS AND "ADDING" VELOCITIES

When considering the relative motion between you and a rocket, I have drawn spacetime diagrams with your worldline lying along the t-axis as if you were stationary. Because of the relativity of velocity, we know that it is equally valid to regard the rocket as stationary, with you moving relative to it. As a consequence, we should be able to re-draw all our spacetime diagrams in such a way that the vertical axis (or time axis) is the worldline of the rocket. (See figure 6.1.) We can also choose the $\hat{x}$-axis to be a line of simultaneity for the rocket pilot. This means that the $\hat{x}$-axis is orthogonal to the $\hat{t}$-axis just as the x-axis is orthogonal to the t-axis. (See the "remark" at the end of the previous chapter for the meaning of "orthogonal.") Each unit along the $\hat{t}$-axis is to be 1 second in the new diagram, and each unit along the $\hat{x}$-axis is to be 1 light-second (186,000 miles). Notice that, since the speed of light is the same relative to the rocket as it is relative to you, photons travel 1 unit in the $\hat{x}$-direction during each unit of $\hat{t}$ time. In this new reference system therefore, the null lines still have slopes of +1 or -1, so they are still the lines which make a 45° angle with the coordinate axes. No matter whose reference system we adopt, the null lines will be the 45° lines.

Let's now recall diagram 4-23, which represents a rocket moving relative

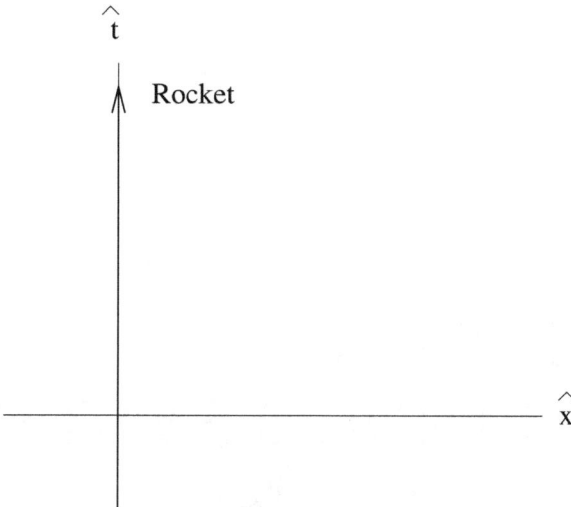

Figure 6.1:

to you with speed v along the line in space which connects the two of you. See figure 6.2. Regarding herself as stationary, the rocket pilot would say

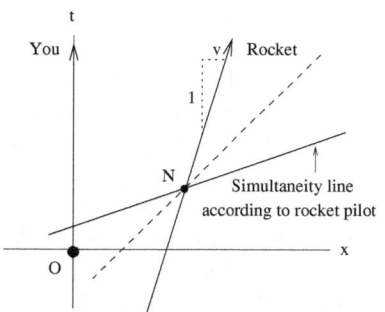

Figure 6.2:

that you are moving with speed v in the other direction. In her reference system your worldline is slanted, as in figure 6.3.

Since photons have 45° worldlines in this new reference system as well, radar measurements look the way they did in chapter 4. By the same reasoning as in that chapter, the events which you would regard as simultaneous with event O can be found by drawing a dotted 45° line through

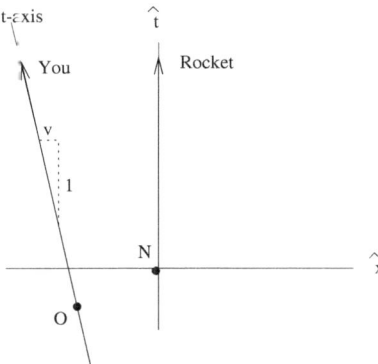

Figure 6.3:

O and finding the line which is symmetrical with your worldline. See figure

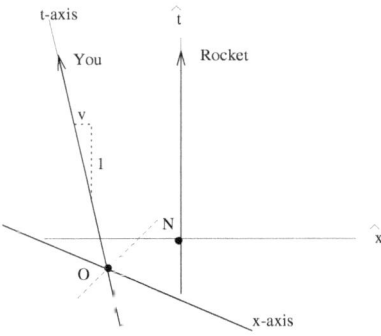

Figure 6.4:

6.4. That is how the old x-axis looks in the new reference system.

To gain familiarity with this switching of reference system, let's convert some more of our diagrams from your reference system to that of the rocket. The left diagram in figure 6.5 is figure 4.4 except that a line orthogonal to the rocket's worldline has been drawn through event B. That line becomes the $\hat{x}$-axis. *Any* line orthogonal to the rocket's worldline could be chosen as the $\hat{x}$-axis. For example, we could choose the line to contain event O as in figure 6.6. Figure 6.7 shows the conversion of figure 4.7, with the $\hat{x}$-axis chosen to pass through event N. Notice that events O and N, which the rocket pilot regards as simultaneous, occur at the same coordinate time $\hat{t}$ in her reference system.

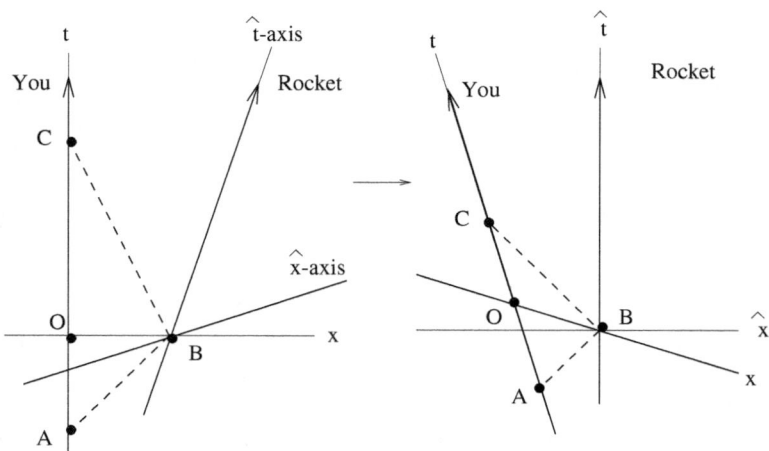

Figure 6.5:

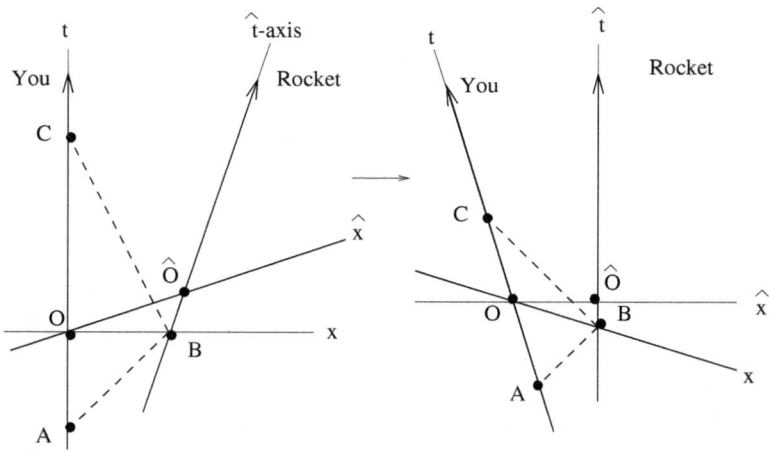

Figure 6.6:

Switching Reference Systems and "Adding" Velocities

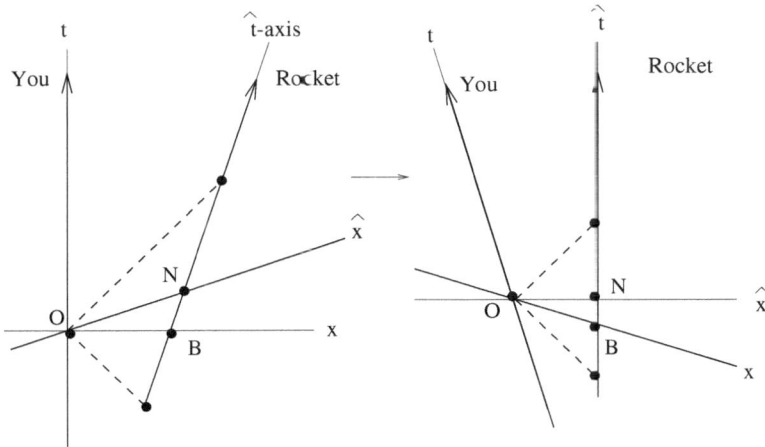

Figure 6.7:

By switching to a different reference system, we are merely choosing a new pair of orthogonal lines to use as coordinate axes. If a photon bounces off you at some event, that event can be identified on your worldline in either picture. In fact any event in one picture can be located in the other picture. An event E in your reference system has some coordinates, particular values for x and t. In the rocket's reference system the same event is labeled by another pair of numbers, these being its $\hat{x}$-coordinate and $\hat{t}$-coordinate. In the two diagrams of figure 6.8, event E has coordinates $(x,t) = (7,6)$ in your reference system and $(\hat{x}, \hat{t}) = (4,2)$ in the rocket pilot's reference system. It is the same event in Minkowski spacetime regardless of which coordinate system we use for labeling it.

For any *pair* of events in Minkowski spacetime, there is an associated number – the interval. That number does *not* depend on how we choose coordinates. We started with a coordinate system (x,t) in which any interval could be calculated by the formula

$$\mathcal{I} = -(x_2 - x_1)^2 + (t_2 - t_1)^2 \tag{6.1}$$

We have now introduced a new coordinate system $(\hat{x}, \hat{t})$. These two coordinate systems have these three properties in common (1) the coordinate axes are orthogonal lines, (2) each unit along the timelike axis is a separation of one second, and (3) each unit along the spacelike axis is a distance of 1 light-second. As you'll see, the interval between events $(\hat{x}_1, \hat{t}_1)$ and $(\hat{x}_2, \hat{t}_2)$ can be computed using the formula

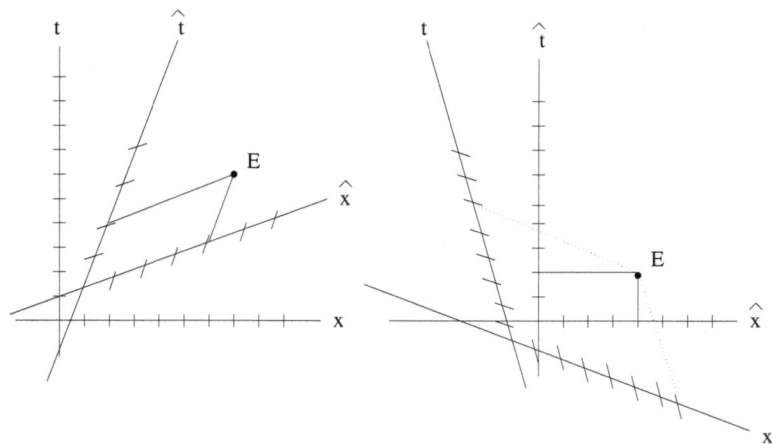

Figure 6.8:

$$\mathcal{I} = -(\hat{x}_2 - \hat{x}_1)^2 + (\hat{t}_2 - \hat{t}_1)^2 \tag{6.2}$$

This is the same as our good old Minkowski interval formula 2.2 except the coordinates $(\hat{x}, \hat{t})$ of the rocket's reference system are used instead of the (x, t) coordinates of your reference system.

There was nothing special about that rocket. No matter what non-accelerating thing you choose, you can regard that thing as stationary and measure all velocities relative to it. You can use a reference system (or coordinate system) in which its worldline is the timelike axis. The spacelike axis is some line orthogonal to the worldline of the thing, and hence a line of events which the thing would regard as simultaneous. If units along the axes are chosen to correspond to seconds and light-seconds, then Minkowski intervals can be computed by the good old Minkowski interval formula, using the coordinates of the thing's reference system if you please.

At the end of the chapter on "The Relativity of Velocity" I asserted that the Minkowski spacetime geometry looks the same to all non-accelerating persons. A more precise form of this assertion is the one we have now arrived at: The formula for Minkowski intervals is the same old Minkowski interval formula no matter what non-accelerating reference system is used. If the formula were *not* the same in any such reference system, then one could stipulate that velocities should be measured relative to a person in whose reference system the interval formula had some special form. However, since the formula is the same in all the reference systems, the Minkowski spacetime geometry does not discriminate between different non-accelerating persons, and anyone has equal right to regard himself as stationary.

It still remains to verify that the Minkowski interval formula does indeed have the same form in a new coordinate system with orthogonal axes, etc. There is a parallel problem for Euclidean space: to verify that the distance formula is given by

$$\mathcal{D}^2 = (x_2 - x_1)^2 + (y_2 - y_1)^2 \tag{6.3}$$

in any coordinate system with (1) orthogonal (i.e. perpendicular) axes, (2) each unit along the x-axis being one unit of distance, and (3) each unit along the y-axis being one unit of distance. In what follows, the left column deals with this problem in Euclidean space; the right column deals with the parallel problem in Minkowski spacetime.

We begin with coordinates (x, y) and distances between points given by

$$\mathcal{D}^2 = (x_2 - x_1)^2 + (y_2 - y_1)^2$$

Consider any line (to become the $\hat{x}$-axis) and suppose its slope is m (sliding along the line, the y-coordinate changes by m as the x-coordinate increases by 1 unit). Pick any perpendicular line (to become the $\hat{y}$-axis), and suppose it meets the $\hat{x}$-axis at the point $\hat{O}$ where $x = a$ and $y = b$. See left sides of figures 6.9 and 6.10.

We begin with coordinates (x, t) and intervals between events given by

$$\mathcal{I}_{EF} = -(x_2 - x_1)^2 + (t_2 - t_1)^2$$

Consider any timelike line (to become the $\hat{t}$-axis) and suppose it represents an object with velocity v relative to a person whose worldline is the t-axis. Pick any orthogonal line (to become the $\hat{x}$-axis), and suppose it meets the $\hat{t}$-axis at the event $\hat{O}$ where $x = a$ and $t = b$. See right sides of figures 6.9 and 6.10.

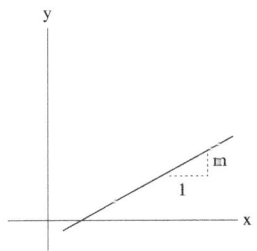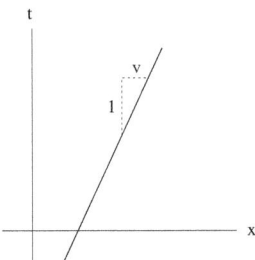

Figure 6.9:

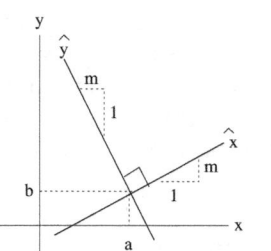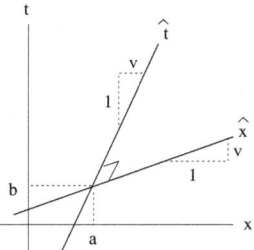

Figure 6.10:

If $(\hat{x}, \hat{y})$ are the coordinates of some arbitrary point with respect to the new system, then that same point has (x, y) coordinates given by

$$x = \frac{1}{\sqrt{1+m^2}}(\hat{x} - m\hat{y}) + a$$
$$y = \frac{1}{\sqrt{1+m^2}}(m\hat{x} + \hat{y}) + b$$

You may check these expressions on the following points whose coordinates in either system can be deduced from the picture (figure 6.11):
$\hat{O} : (\hat{x}, \hat{y}) = (0, 0), (x, y) = (a, b)$
$R : (\hat{x}, \hat{y}) = (\sqrt{1+m^2}, 0), (x, y) = (a+1, b+m)$
$S : (\hat{x}, \hat{y}) = (0, \sqrt{1+m^2}), (x, y) = (a-m, b+1)$

If $(\hat{x}, \hat{t})$ are the coordinates of some arbitrary event with respect to the new system, then that same event has (x, t) coordinates given by

$$x = \frac{1}{\sqrt{1-v^2}}(\hat{x} + v\hat{t}) + a$$
$$t = \frac{1}{\sqrt{1-v^2}}(v\hat{x} + \hat{t}) + b \quad (6.4)$$

You may check these expressions on the following events whose coordinates in either system can be deduced from the picture (figure 6.11):
$\hat{O} : (\hat{x}, \hat{t}) = (0, 0), (x, t) = (a, b)$
$G : (\hat{x}, \hat{t}) = (\sqrt{1-v^2}, 0), (x, t) = (a+1, b+v)$
$H : (\hat{x}, \hat{t}) = (0, \sqrt{1-v^2}), (x, t) = (a+v, b+1)$

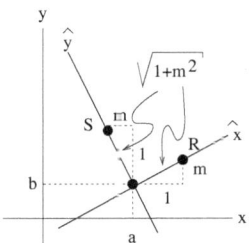
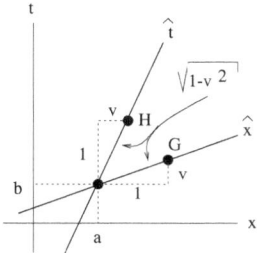

Figure 6.11:

Let P and Q be two points. In the (x, y) coordinate system suppose P has coordinates (x_1, y_1) and Q has coordinates (x_2, y_2). We know that the squared distance $\mathcal{D}^2_{PQ}$ can be computed from the formula

$$\mathcal{D}^2_{PQ} = (x_2 - x_1)^2 + (y_2 - y_1)^2$$

With respect to the $(\hat{x}, \hat{y})$ system the events P and Q have coordinates $(\hat{x}_1, \hat{y}_1)$ and $(\hat{x}_2, \hat{y}_2)$. These numbers are related to the (x, y) coordinates by

$$x_2 = \frac{1}{\sqrt{1+m^2}}(\hat{x}_2 - m\hat{y}_2) + a$$

$$x_1 = \frac{1}{\sqrt{1+m^2}}(\hat{x}_1 - m\hat{y}_1) + a$$

$$y_2 = \frac{1}{\sqrt{1+m^2}}(m\hat{x}_2 + \hat{y}_2) + b$$

$$y_1 = \frac{1}{\sqrt{1+m^2}}(m\hat{x}_1 + \hat{y}_1) + b$$

If you now substitute the right hand sides of these equations for the numbers x_2, x_1, y_2, and y_1 in the calculation of $\mathcal{D}^2_{PQ}$, you'll find that it simplifies to

$$\mathcal{D}^2_{PQ} = (\hat{x}_2 - \hat{x}_1)^2 + (\hat{y}_2 - \hat{y}_1)^2$$

This verifies that the formula for computing distances like $\mathcal{D}_{PQ}$ is the same in the $(\hat{x}, \hat{y})$ coordinate system as in the (x, y) system.

Let E and F be two events. In the (x, t) coordinate system suppose E has coordinates (x_1, t_1) and F has coordinates (x_2, t_2). We know that the interval can be computed from the formula

$$\mathcal{I}_{EF} = -(x_2 - x_1)^2 + (t_2 - t_1)^2$$

With respect to the $(\hat{x}, \hat{t})$ system the events E and F have coordinates $(\hat{x}_1, \hat{t}_1)$ and $(\hat{x}_2, \hat{t}_2)$. These numbers are related to the (x, t) coordinates by

$$x = \frac{1}{\sqrt{1-v^2}}(\hat{x} + v\hat{t}_2) + a$$

$$x_1 = \frac{1}{\sqrt{1-v^2}}(\hat{x}_1 + v\hat{t}_1) + a$$

$$t_2 = \frac{1}{\sqrt{1-v^2}}(v\hat{x}_2 + \hat{t}_2) + b$$

$$t_1 = \frac{1}{\sqrt{1-v^2}}(v\hat{x}_1 + \hat{t}_1) + b$$

If you now substitute the right hand sides of these equations for the numbers x_2, x_1, t_2, and t_1 in the calculation of $\mathcal{I}_{EF}$, you'll find that it simplifies to

$$\mathcal{I}_{EF} = -(\hat{x}_2 - \hat{x}_1)^2 + (\hat{t}_2 - \hat{t}_1)^2$$

This verifies that the formula for computing intervals like $\mathcal{I}_{EF}$ is the same in the $(\hat{x}, \hat{t})$ coordinate system as in the (x, t) system.

The $(\hat{x}, \hat{y})$ system was not special among coordinate systems satisfying the three conditions. Any such system can be specified by the slope m of its $\hat{x}$-axis (relative to the xy-system) and the coordinates $(x = a, y = b)$ of the point where its axes meet. Since we have done the calculation for arbitrary values of m, a, and b, we now know that the distance formula is the same in any such coordinate system.

The $(\hat{x}, \hat{t})$ system was not special among coordinate systems satisfying the three conditions. Any such system can be specified by the velocity v of the $\hat{t}$-axis (relative to the t-axis) and the coordinates $(x = a, y = b)$ of the event where its axes meet. Since we have done the calculation for arbitrary values of v, a, and b (but $|v| < 1$), we now know that the interval formula is the same in any such coordinate system.

Recreation: Suppose a rocket is moving at $\frac{3}{5}$ the speed of light relative to you. The origin of coordinates $\hat{O}$ in the rocket's reference system is to be the event (1,1) in your coordinate system. See figure 6.12. Let E be the event whose $(\hat{x}, \hat{t})$ coordinates are (1,2). Use the formulae 6.4 to find the coordinates of E in your (x, t) coordinate system. (Answer: $(\frac{15}{4}, \frac{17}{4})$) Use the Minkowski interval formula 2.2 to show that the separation between $\hat{O}$ and E is equal to $\sqrt{3}$. Then use the formula 6.2 in the $(\hat{x}, \hat{t})$ system to compute the separation of $\hat{O}$ and E. You must get the same answer: $\sqrt{3}$.

Velocity "Addition"

Imagine for a moment that your mother is going on an airplane journey and you have taken her to the airport. Your farewell takes place as she steps onto a speed ramp which will take her to the departure gate. The moving walkway will carry your mother at a speed of 4 miles per hour as she stands and waves good-bye. Just as she steps onto the ramp, a military officer also steps on, but he walks along the walkway at his customary walking speed of 3 miles per hour. Question: How fast is the officer moving relative to you as you stand and wave? You might answer that the officer is moving 7 miles per hour relative to you. After all, he is moving 3 miles per hour relative to the ramp, and the ramp is moving 4 miles per hour relative to you. This answer, however, is not completely accurate, and that method of adding velocities is grossly inaccurate for speeds which are close to the speed of light. For example, suppose the speed ramp moves at $\frac{3}{4}$ the speed of light relative to you, and suppose the officer walks at $\frac{3}{4}$ the speed of light relative to the ramp. If one were to add the two speeds he would get a

Switching Reference Systems and "Adding" Velocities

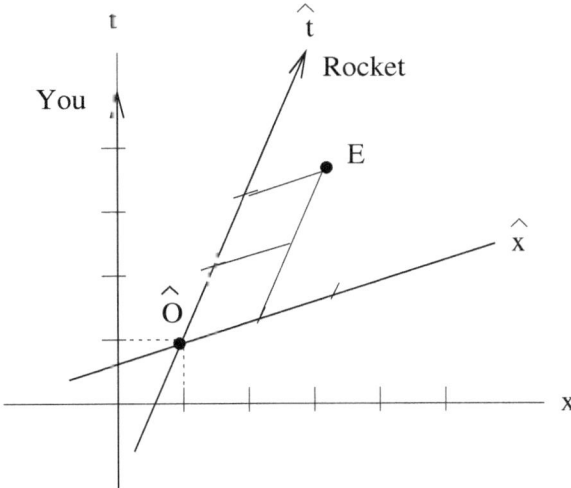

Figure 6.12:

speed of $1\frac{1}{2}$ times the speed of light. But it is impossible for the officer to move relative to you faster than the speed of light! It is incorrect simply to add the two relative velocities.

Let's figure out exactly how fast the officer is moving relative to you if your mother has speed v relative to you and the officer is walking with speed w relative to your mother. One way to find the answer is to use the formulae 6.4 for changing coordinate systems. In your mother's reference system, the situation looks like figure 6.13.

She is stationary. You and the officer move away from her in opposite directions. After 1 second has elapsed on her clock, you are at a distance v and the officer is at a distance w from her. To find the officer's speed relative to you, we can switch to your reference system. The officer's speed relative to you is easy to compute once we know the coordinates of event G. If (p, q) are the coordinates of event G, then $s = \frac{p}{q}$ is the officer's speed relative to you, because he goes those p light-seconds during the q seconds on your clock. In the $(\hat{x}, \hat{t})$ coordinate system of your mother, the coordinates of event G are $(w, 1)$, as shown in figure 6.13. You can use the formulae 6.4, with $a = 0, b = 0, \hat{x} = w$, and $\hat{t} = 1$, to find the (x, t) coordinates of G. (Your mother is moving with speed v relative to you.) You'll find

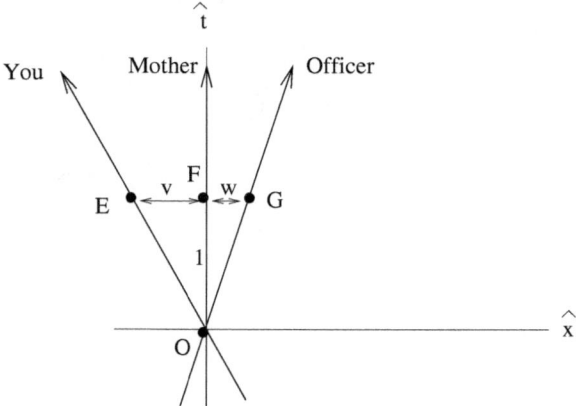

Figure 6.13:

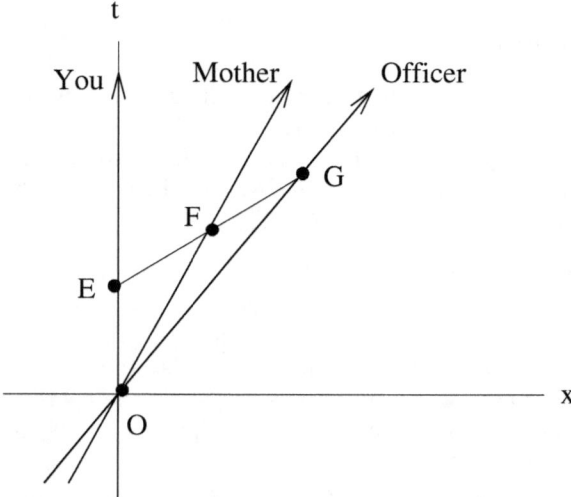

Figure 6.14:

$$p = \frac{v+w}{\sqrt{1-v^2}} \quad \text{and} \quad q = \frac{1+vw}{\sqrt{1-v^2}} \tag{6.5}$$

The speed $(s = \frac{p}{q})$ is therefore

$$s = \frac{v+w}{1+vw} \tag{6.6}$$

This boxed formula gives the correct way to "add" relative velocities. Notice that, if v and w are very small fractions, the product vw is extremely tiny and s is quite accurately given by $v+w$. Since speeds of 3 and 4 miles per hour are very small fractions of the speed of light, it is quite accurate (but not exact) to say that the officer is moving 7 miles per hour relative to you in answer to the question first posed. If $v = \frac{3}{4}$ and $w = \frac{3}{4}$, however, you will see the officer moving with speed

$$s = \frac{\frac{3}{4}+\frac{3}{4}}{1+(\frac{3}{4})(\frac{3}{4})} = \frac{24}{25}$$

This speed is less than the speed of light. In fact, for any fractions v and w (with absolute values between 0 and 1) the combined velocity s will also have absolute value between 0 and 1.

Remark: Switching reference systems was a *convenience* in solving this problem. It was not *necessary*. In fact, it is never necessary to switch reference systems. We could choose once and for all a particular reference system and then stick with it. To illustrate this, I'd like to derive the boxed formula for velocity "addition" using only the reference system in which your worldline is the t-axis, as in figure 6.15. The objective, as before, is to find a way to express the coordinates (p, q) of event G in terms of the speeds v and w. The combined speed s will then be given by $s = \frac{p}{q}$.

Let's first find an expression for p. As shown in the figure, let H be the event which you regard as simultaneous with G. Because the line HG is orthogonal to the line EH, we know that

$$\mathcal{I}_{EG} = \mathcal{I}_{HG} + \mathcal{I}_{EH} \tag{6.7}$$

(See the remark at the end of chapter 5.) We know, moreover, that $\overline{EG} = v + w$, so $\mathcal{I}_{EG} = -(v+w)^2$. Also, the line EF has slope v since it is a line of simultaneity for your mother moving with speed v relative to you. Its slope being v means that $\overline{EH} = v \cdot \overline{HG}$. Now $\overline{HG}$ is just p, so $\overline{EH} = vp$. We can write $\mathcal{I}_{HG} = -p^2$ and $\mathcal{I}_{EH} = v^2 p^2$. Substituting all these expressions into equation 6.7 yields

$$-(v+w)^2 = -p^2 + v^2 p^2$$

You can solve this for the positive number p:

$$p = \frac{v+w}{\sqrt{1-v^2}}$$

Now for q, the t-coordinate of event G. We can write $q = \overline{OE} + \overline{EH}$. We have already observed that $\overline{EH} = vp$, and so $\overline{EH} = v \frac{v+w}{\sqrt{1-v^2}}$. What about $\overline{OE}$? Well, the line EF is orthogonal to your mother's worldline OF, so

$$\mathcal{I}_{OE} = \mathcal{I}_{EF} + \mathcal{I}_{OF} \tag{6.8}$$

We know that $\mathcal{I}_{EF} = -v^2$ and $\mathcal{I}_{OF} = 1$, and so the above equation becomes $\mathcal{I}_{OE} = -v^2 + 1$. The separation $\overline{OE}$ is therefore $\sqrt{1-v^2}$. We can now calculate $q = \overline{OE} + \overline{EH}$:

$$q = \sqrt{1-v^2} + v \frac{v+w}{\sqrt{1-v^2}} = \frac{1+vw}{\sqrt{1-v^2}}$$

The quotient $s = \frac{p}{q}$ agrees with the boxed formula.

Switching Reference Systems and "Adding" Velocities

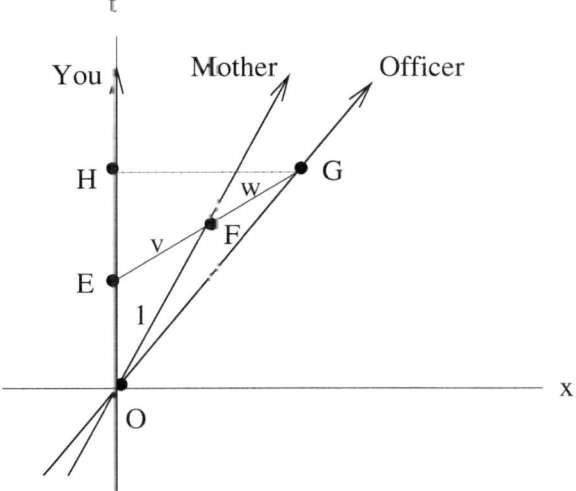

Figure 6.15:

In 4-dimensional Minkowski spacetime. If you and a rocket have a relative speed v along the x-direction, then the y-coordinates and z-coordinates of all events may be the same in both reference systems. In other words, the coordinates (x, y, z, t) of your reference system can be related to the $(\hat{x}, \hat{y}, \hat{z}, \hat{t})$ coordinates of the rocket's system by the formulae

$$\begin{aligned} x &= \frac{1}{\sqrt{1-v^2}}(\hat{x} + v\hat{t}) + a \\ y &= \hat{y} \\ z &= \hat{z} \\ t &= \frac{1}{\sqrt{1-v^2}}(v\hat{x} + \hat{t}) + b \end{aligned} \qquad (6.9)$$

Let's consider a velocity "addition" problem which cannot be confined to the xt-plane: Imagine that you are going on an airplane journey, and your father has taken you to the airport. After saying farewell, you step onto a speed ramp which moves you toward the west at 4 miles per hour. As you are stepping onto the ramp, a flight attendant passes you and your father. She is walking north at 3 miles per hour across the lobby floor. Question: How fast is the flight attendant moving relative to you as you stand and wave good-bye on the moving walkway?

Let me first suggest an answer which you might give if you had never heard of Albert Einstein or Minkowski spacetime. You could reason that, after one hour, you would be 4 miles west of where your father stood, and the flight attendant would be 3 miles north of that spot. The (Euclidean) distance between you and the flight attendant would then be

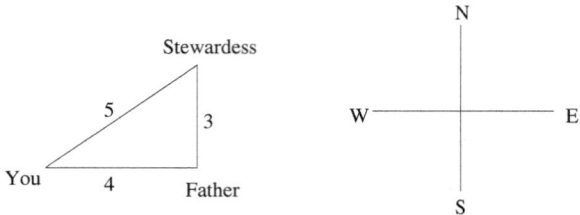

Figure 6.16:

$\mathcal{D} = \sqrt{4^2 + 3^2} = 5$; i.e. after one hour the flight attendant would be 5 miles away from you. That would mean that she were moving 5 miles per hour relative to you. This answer is not completely correct, however, and the method would lead to serious errors if the speeds in question were close to the speed of light.

To answer the question correctly, it is convenient to begin with your father's reference system. Suppose O is the event where you step on the speed ramp and the flight attendant walks by. Let this event have coordinates (0,0,0,0) in both your father's coordinate system and your own. For the sake of generality, let's say the speed ramp moves with speed v relative to your father, and the flight attendant walks with speed w relative to him. After 1 second on your father's watch, the flight attendant is at an event G with coordinates $(0, w, 0, 1)$ in your father's $(\hat{x}, \hat{y}, \hat{z}, \hat{t})$ system (where his $\hat{x}$-axis points east and his $\hat{y}$-axis north). Using the formulae 6.8, you may check that event G has coordinates $(\frac{v}{\sqrt{1-v^2}}, w, 0, \frac{1}{\sqrt{1-v^2}})$ in your (x, y, z, t) system. This event is simultaneous, according to you, with the event H on your worldline with coordinates $(x, y, z, t) = (0, 0, 0, \frac{1}{\sqrt{1-v^2}})$. You can verify that the spacelike distance $\overline{HG}$ is equal to

$$\frac{\sqrt{v^2 + w^2 - v^2 w^2}}{\sqrt{1 - v^2}}$$

From your point of view, the flight attendant has moved that distance from you during the timespan of $\frac{1}{\sqrt{1-v^2}}$ seconds. Relative to you, her speed is therefore

$$\sqrt{v^2 + w^2 - v^2 w^2} \tag{6.10}$$

This is the correct answer. If v and w are both very small fractions of the speed of light, then v^2w^2 is negligible compared to $v^2 + w^2$. In that case, it is a good approximation to say her speed is $\sqrt{v^2 + w^2}$ relative to you. That approximate answer agrees with the one considered in the previous paragraph. Note that even if v and w are very close to 1, this combination of velocities remains less than 1. (*Hint:* $v^2 + (1-v^2) \equiv 1$, and since $w^2 < 1$, it must be that $v^2 + w^2(1 - v^2) < 1$. The square root of that must then also be less than 1, so $\sqrt{v^2 + w^2 - v^2w^2} < 1$.)

Chapter 7

TIME DILATION

A moving clock runs slow. Very crudely, this is what is meant by time dilation.

One trouble with this crude statement of time dilation is that it ignores what we know about the relativity of velocity. What is a moving clock? Moving relative to whom? It would be better to say something like this: "If a rocket is moving relative to you, then you will observe the rocket's clock to be running slower than your own clock."

Time dilation is for real. We don't notice it at our everyday speeds, but it is significant when you observe a clock moving at nearly the speed of light relative to yourself. I'd like to cite such a situation where time dilation is important. When cosmic rays from outer space collide with atoms in the Earth's upper atmosphere, they produce a lot of particles called muons. A muon serves as a clock because it decays into an electron and a neutrino after about 2 microseconds, i.e. it breaks apart about two millionths of a second after it is created. In such a short time, even a photon of light would travel only about 2000 feet. A muon cannot travel as fast as light, so it could not even go 2000 feet during a span of 2×10^{-6} second on a ground based clock. Nevertheless, most of the energetic muons produced in the upper atmosphere by cosmic rays do not decay until they have traveled many miles from the upper atmosphere and have penetrated the Earth's surface. To somebody on the ground, the muons seem to live at least a hundred times longer than their typical lifespan. This is due to time dilation. Since the muon moves at nearly the speed of light relative to the ground, a person on the ground observes the muon clock to run very slow compared to his own. While the slow muon clock measures off its lifespan of 2×10^{-6} second, the ground based clock records a timespan more than a hundred times as long. And that's plenty of time for the muon to get from the upper atmosphere to the ground.

I have still not said precisely what is meant by time dilation because I haven't explained exactly what it means for you to "observe a rocket's clock to be running slower than your own clock." In fact, the whole business may seem paradoxical so far. After all, from the rocket pilot's point of view, you and your clock are moving relative to her, and so the rocket pilot should observe your clock to be running slow relative to her own. You observe hers to run slow and she observes yours to run slow! It sounds impossible, but you'll see that it is all consistent once I explain what it means for you to "observe the rocket's clock to be running slower than your own clock." A spacetime diagram will make it easy to be precise about this.

Suppose a rocket goes by you at some event O and the rocket is moving with speed v relative to you. Suppose also that you and the rocket pilot set your clocks to show zero when you are together at event O. Let E be an event on your worldline which occurs τ seconds[1] after event O, and call F the event on the rocket's worldline which you can regard as simultaneous

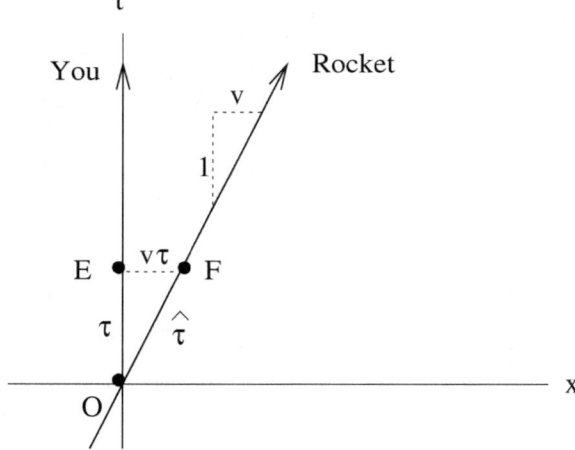

Figure 7.1:

with E. See figure 7.1. At event E your clock shows the quantity $\overline{OE}$, which is just τ. Let's say event O has coordinates $(0,0)$, so E has coordinates $(0, \tau)$. At event F, the clock on the rocket shows $\overline{OF}$. Let $\hat{\tau}$ denote this separation $\overline{OF}$. Since the rocket moves $v\tau$ units in the x-direction as the t-coordinate changes from 0 to τ, the coordinates for event F are $(v\tau, \tau)$; and you can compute the separation $\overline{OF}$ as in chapter 2. You'll find that

---

[1] The symbol τ is the Greek letter "tau." It here stands for any old number of seconds.

Time Dilation

$$\hat{\tau} = \tau\sqrt{1-v^2} \qquad (7.1)$$

At the instant when your clock shows time τ, you would say the rocket's clock shows the smaller value $\tau\sqrt{1-v^2}$. In other words, you can observe the rocket's clock to be running slower than yours.

The boxed equation above is the quantitative statement of time dilation. The spacetime diagram should be remembered along with the formula in order to avoid confusion about the meaning of the quantities τ and $\hat{\tau}$. Notice that, when the speed of the rocket relative to you is small compared to the speed of light $(v \approx 0)$,[2] the timespans τ and $\hat{\tau}$ are just about the same because $\sqrt{1-v^2} \approx 1$. That's why we don't notice time dilation in everyday life. If v is very close to 1 (so the rocket is moving at nearly the speed of light relative to you), then the quantity $\sqrt{1-v^2}$ is a very tiny fraction, and $\hat{\tau}$ is much much smaller than τ.

The rocket pilot sees matters somewhat differently. She would not agree that events E and F in the preceding diagram are simultaneous, so she would see no reason to compare the clocks at those two events. By drawing the line of events which the rocket pilot regards as simultaneous with F, you see that the line meets your worldline at some event G (figure 7.2).

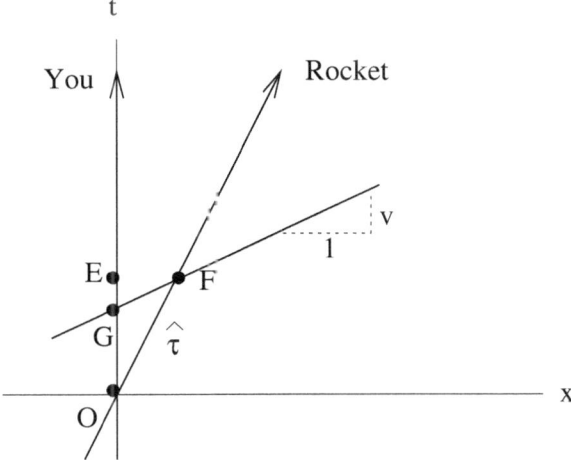

Figure 7.2:

The rocket pilot would compare $\overline{OG}$ to $\overline{OF}$ to find out how much time

---

[2]The symbol $\approx$ means "nearly equals."

has elapsed on your clock while her clock recorded the amount $\hat{\tau}$. Let's compute $\overline{OG}$. The line GF has slope v and contains event F with coordinates $(v\tau, \tau)$. You can check that its equation must be $t = vx + \tau(1 - v^2)$. The x-coordinate of G is zero, and its t-coordinate is the t-intercept of the line GF, namely $\tau(1 - v^2)$. so G has coordinates $(0, \tau(1 - v^2))$, and $\overline{OG} = \tau(1 - v^2)$. This can be written as:

$$\overline{OG} = \hat{\tau}\sqrt{1 - v^2} \tag{7.2}$$

(since $\hat{\tau} = \tau\sqrt{1 - v^2}$). When the rocket is at event F, the pilot would say her clock shows $\hat{\tau}$ simultaneously with your clock showing the smaller value $\hat{\tau}\sqrt{1 - v^2}$. She observes your clock to be running slower than hers.

This observation by the rocket pilot is perhaps easier to discover if we use her reference system (figure 7.3). In her $(\hat{x}, \hat{t})$ coordinates, event F

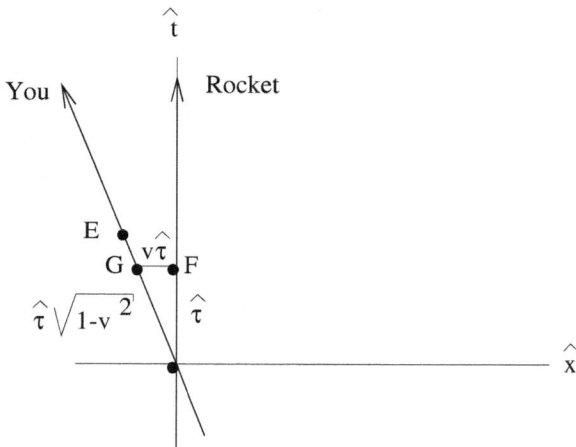

Figure 7.3:

has coordinates $(0, \hat{\tau})$. Since G is simultaneous with F in this system, the $\hat{t}$-coordinate of G is $\hat{\tau}$. And since she sees you move $v\hat{\tau}$ units in the negative $\hat{x}$-direction during the time $\hat{\tau}$, the event G must have coordinates $(-v\hat{\tau}, \hat{\tau})$. We can use the good old Minkowski interval formula with her coordinates and easily check that

$$\overline{OG} = \hat{\tau}\sqrt{1 - v^2} \tag{7.3}$$

so we again find that she sees your clock as slow.

Time Dilation

> *Recreation*: Since you regard events E and F as simultaneous, let's again compare the separations $\overline{OE}$ and $\overline{OF}$, this time using the rocket's coordinate system for doing the calculation. The separation $\overline{OF}$ is just $\hat{\tau}$. What about $\overline{OE}$? Well, E is the unique event shared by the lines OE and EF. The line OE is your worldline and has equation
>
> $$\hat{t} = -\frac{1}{v}\hat{x} \tag{7.4}$$
>
> The line EF is a line of events which you regard as simultaneous. Show that its equation is
>
> $$\hat{t} = -v\hat{x} + \hat{\tau} \tag{7.5}$$
>
> Verify that the one event common to these two lines (i.e. event E) has coordinates
>
> $$(-\frac{v\hat{\tau}}{1-v^2}, \frac{\hat{\tau}}{1-v^2})$$
>
> Check that $\overline{OF} = \overline{OE}\sqrt{1-v^2}$. This shows that you observe the rocket's clock to run slow relative to your own.

In summary: If you and a rocket have a relative speed v, you observe the rocket's clock to record only $\tau\sqrt{1-v^2}$ seconds while your clock ticks away τ seconds. Similarly, the rocket pilot would say that your clock advances only $\hat{\tau}\sqrt{1-v^2}$ seconds while hers advances $\hat{\tau}$ seconds. Each observes the other's clock to run slow. There is no inconsistency because you and she do not agree about which events are simultaneous, and so she does not compare the same separations which you compare. In the next chapter we'll consider a circumstance in which the clocks come back together for an unambiguous comparison.

In 4-dimensional Minkowski spacetime. The time dilation effect is the same as in 2-dimensional spacetime of the xt-plane. If a rocket is moving with speed v relative to you, then you will say that its clock measures only $\tau\sqrt{1-v^2}$ seconds while your clock measure τ seconds. It does not matter whether or not the rocket's worldline meets your worldline. (If it does not, then the rocket is moving neither directly at you nor directly away from you.) You can imagine that you have a buddy who is stationary relative to you and that the rocket passes right by your buddy. Think of the xt-plane as the plane containing the crossing worldlines of your buddy and the rocket. Since your buddy is motionless compared to you, the rocket moves with speed v also relative to him. Moreover, he keeps the same time as you and agrees with you about simultaneity. We know that he would observe the rocket's clock to advance only $\tau\sqrt{1-v^2}$ seconds while his advances τ seconds, and you would observe the same thing.

Chapter 8

THE TWIN EFFECT

Alice and Billy are twins, born in space on a station not far from Earth. Just after birth Billy begins a journey to the other side of our Milky Way galaxy and back. He has the very best equipment and can travel at nearly the speed of light. Nevertheless, it is a long journey, one which takes a light signal 100,000 years to complete. When Billy returns to the space station, his twin sister Alice is dead and no longer remembered. It is more than 100,000 years later on the space station. Billy, however, is still alive and looking forward to retirement. This is the twin effect – a dramatic feature of relativity.

The twin effect is a consequence of time dilation. As Alice and her descendants grow old and die, Billy's clock is running extremely slow, and so Billy remains young.

The effect can *seen* paradoxical by arguing that if Billy regards himself as being at rest, then he sees Alice always moving at high speed and it should be *her* clock which runs slow. The twin effect has sometimes been called the "twin paradox." In fact there is no paradox. Alice is correct in calculating Billy's age using the simple time dilation formula 7.1. Billy, however, needs to be more careful *because he reverses his direction of motion* at the far end of his journey. It is Billy's turning around which causes the twins' lives to be unsymmetrical and which forces Billy to reckon Alice's age more carefully.

Some spacetime diagrams may help to make it clear. Regarding Alice as being at rest, we can draw the worldlines as in figure 8.1.
Billy leaves at O, reverses direction at F, and arrives back at the space station at event G. Let me also sketch (figure 8.2 how it looks to a person who is sitting with Billy on the outbound journey, but who never undergoes accelerations (as Billy does at event F):

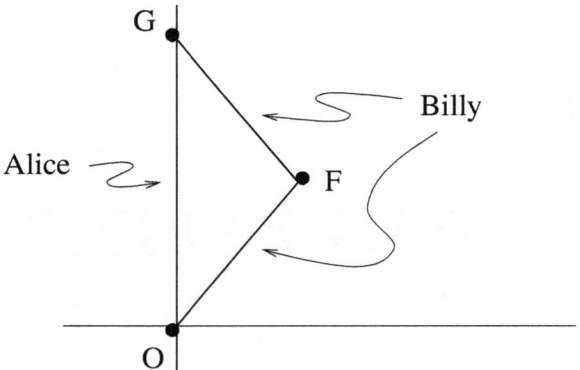

Figure 8.1:

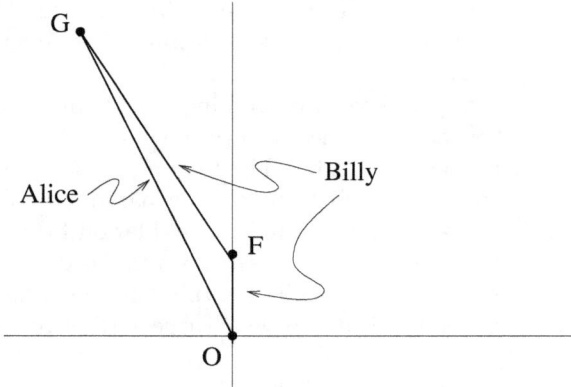

Figure 8.2:

The Twin Effect

And figure 8.3 is how it looks in the reference system of a non-accelerating person whose worldline is shared by Billy during Billy's *return* flight.

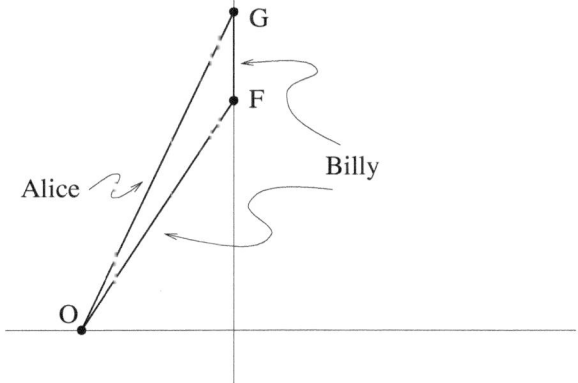

Figure 8.3:

In all views it is Billy's worldline which is kinky. Alice's worldline is straight.

What are the possible ways for a person to get from event O to event G? The worldlines of Alice and Billy represent just two possibilities. The primary restriction is that the person's speed be less than that of light. This means that at any event of the person's worldline, the tangent line at that event must be a timelike line. An allowable worldline of this type is said to be everywhere timelike. The first diagram in figure 8.4 shows an allowable worldline, but the worldline in the right diagram is not everywhere timelike. (At a kink, as in Billy's worldline, both tangent lines should be timelike so

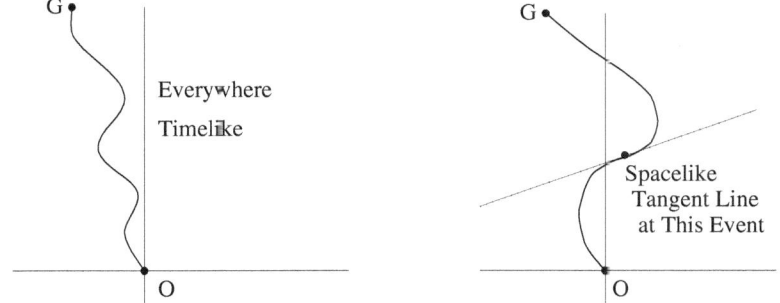

Figure 8.4:

that the speed is less than that of light before and after the kink.)

Here is the relevance of all this: Whereas in Euclidean space the shortest path joining two points is the straight line segment, the situation is radically different in Minkowski spacetime. Among all paths which are everywhere timelike and join events O and G (timelike-separated), the *longest* path is the straight line segment from O to G. Let me try to convince you of this. The diagram in figure 8.5 shows a path from O to G which has *zero* length because each segment is null. The diagram in figure 8.6 exhibits a wiggly

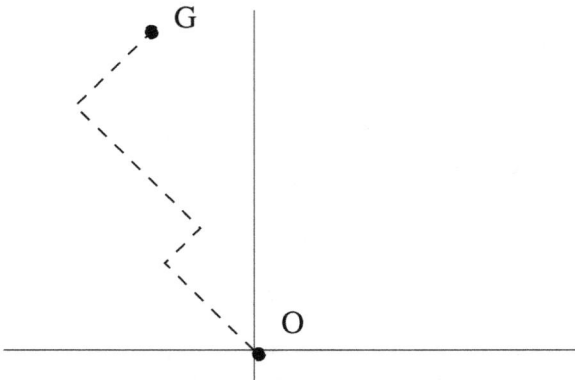

Figure 8.5:

timelike path from O to G which is not much different from the kinky null path of figure 8.5, so the length of this wiggly path is not much different from zero. If you take a straight timelike line segment and deform it so

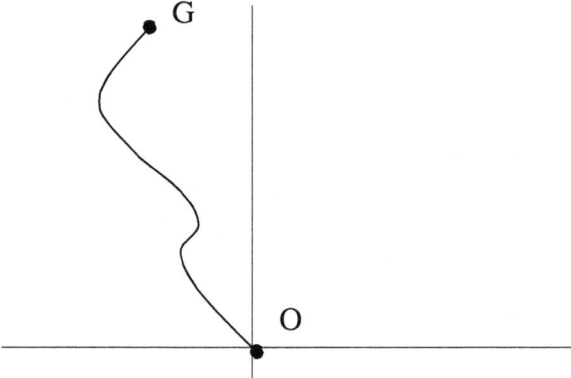

Figure 8.6:

that it has wiggles or kinks (but the same two end-events), then parts of the path are going to be more nearly null, and the deformed path will have shorter length than the straight line segment.

This, then, is a direct way to see that Alice is older than Billy when Billy returns. Alice followed the longest possible worldline from O to G. It doesn't matter which non-accelerating reference system you adopt, it is Alice's worldline which is the straight one.

Let's now be more quantitative and figure out just how much older Alice is when the twins meet again at event G. We can do it using Alice's

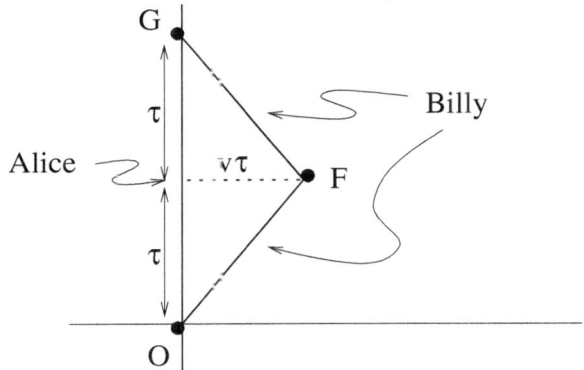

Figure 8.7:

reference system if we please (figure 8.7).

Suppose Alice's clock measures a time span of 2τ seconds between O and G. If O has coordinates $(0,0)$, then G has coordinates $(0, 2\tau)$, and F has coordinates $(v\tau, \tau)$ since Billy travels $v\tau$ light-seconds during τ seconds. At event G, Billy's clock will show a time equal to $\overline{OF} + \overline{FG}$. Calculating the separations as in chapter 2, you'll find that $\overline{OF}$ and $\overline{FG}$ are both equal to $\tau\sqrt{1-v^2}$. At event G, then, billy's clock shows $2\tau\sqrt{1-v^2}$ compared with the greater time 2τ on Alice's clock. Billy is therefore younger than Alice by the factor $\sqrt{1-v^2}$ when they meet at event G.

Remark: It is only a little bit more difficult to do this calculation in the reference system of somebody traveling with Billy on the outbound journey (figure 8.8). Suppose O has coordinates $(0,0)$ and event F has coordinates $(0, \tau)$. What are the coordinates of event G? Event G is the unique event contained in the line OG as well as the line FG. If we write the equations for these two lines, we can find the unique coordinate pair $(\hat{x}, \hat{t})$ which satisfies both equations, and those will be the coordinates for G. Since Billy and Alice have a relative speed of v, the line OG has slope $-\frac{1}{v}$ and its equation is

$$\hat{t} = -\frac{1}{v}\hat{x} \tag{8.1}$$

To find the slope of the line FG, notice that it is the worldline of Billy. Billy has a speed of v relative to Alice, and Alice has a speed v relative to the person whose worldline is the $\hat{t}$-axis. By correctly "adding" the velocities as in chapter 6, we find that Billy, when returning, has a speed

$$s = \frac{2v}{1+v^2} \tag{8.2}$$

relative to the $\hat{t}$-axis. So the slope of line FG is $-\frac{1+v^2}{2v}$. Since the $\hat{t}$-intercept of this line is $\hat{\tau}$, its equation is

$$\hat{t} = -\frac{1+v^2}{2v}\hat{x} + \hat{\tau} \tag{8.3}$$

If you solve equations 8.1 and 8.3 together, you'll find that the solution $(\hat{x}, \hat{t})$ is

$$(-2\frac{v\hat{\tau}}{1-v^2}, 2\frac{\hat{\tau}}{1-v^2})$$

These are the coordinates of event G. Now you can compute the separations:

$$\overline{OG} = \frac{2\hat{\tau}}{\sqrt{1-v^2}}, \quad \overline{OF} = \hat{\tau}, \quad \overline{FG} = \hat{\tau} \tag{8.4}$$

At G Billy's clock shows $\overline{OF} + \overline{FG}$, which is $2\hat{\tau}$. Alice's clock shows the greater time $\frac{2\hat{\tau}}{\sqrt{1-v^2}}$. We have the same conclusion as before: Billy is younger at G by the factor $\sqrt{1-v^2}$.

The Twin Effect

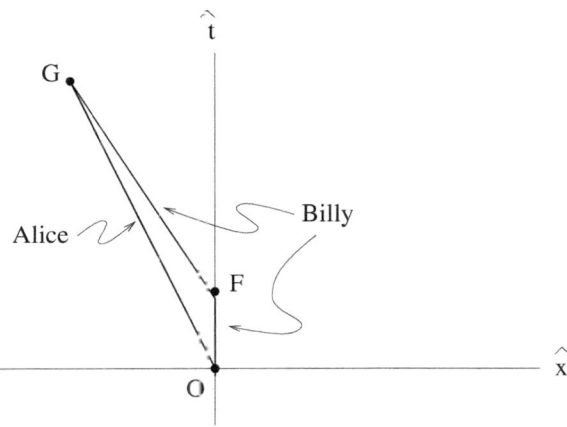

Figure 8.8:

Recreation: Check that you arrive at the same conclusion if you do the calculation in the reference system of figure 8-3.

Recreation: Suppose Alice and Billy both make space trips, going off in opposite directions. If each moves with speed v relative to the space station and both turn around at the same time (according to, say, the space station clock), are they the same age when they arrive back at the space station? See figure 8.9. How have they aged relative to their friends who stayed at the space station?

Recreation: Suppose Billy's spacecraft fails to turn around, so Alice goes to rescue him (figure 8.10). Who is older when she catches up with him?

I think it's illuminating to examine in more detail the way Billy and Alice see each other age. Let's consider once again the original scenario in which Billy travels out and back while Alice remains on the non-accelerating space station. A good way for the twins to keep track of each other is for each to emit one flash of light every second (or every heartbeat if you prefer). By watching for the light flashes, each twin can monitor the other's aging. Let's take figure 8.1 and put in some of the light flashes from Alice to find out what Billy sees. Notice (figure 8.11) that Billy receives relatively few flashes while outbound, but he starts collecting them rapidly as soon as he

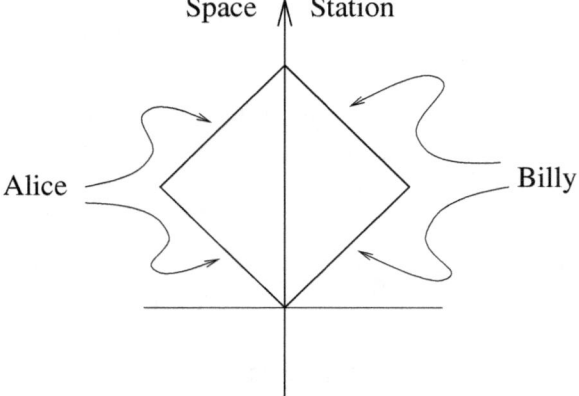

Figure 8.9:

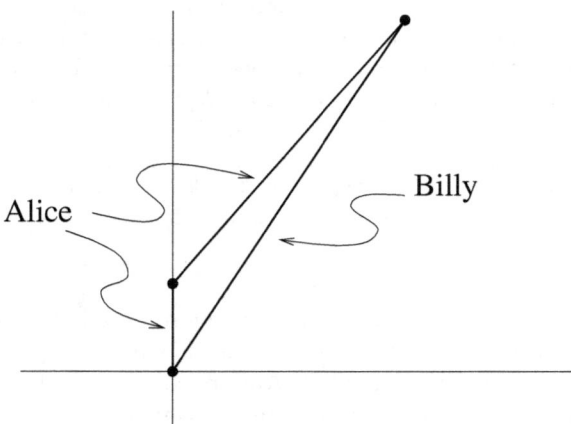

Figure 8.10:

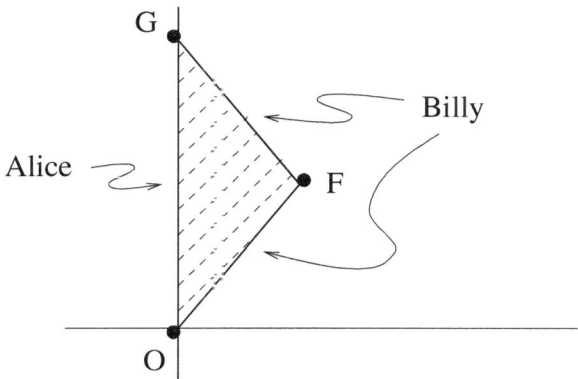

Figure 8.11:

reverses his motion at event F.

I'd like you to consider a question whose answer will be helpful to us. If Alice emits 1 flash per second, at what rate does Billy receive flashes while outward bound? The answer can be deduced by reference to figure 8.12. The total time $\overline{OG}$ of Billy's journey as measured by Alice's clock is 2τ, so

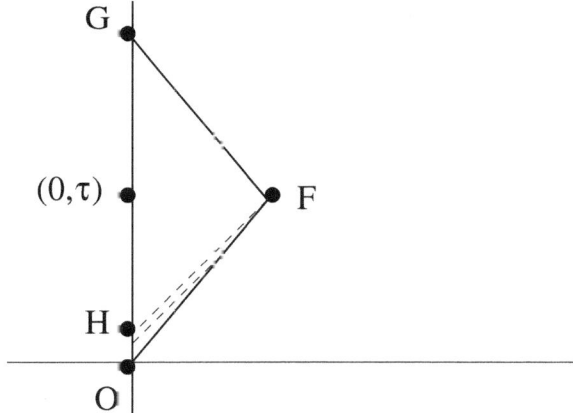

Figure 8.12:

$(0, \tau)$ is the event on her worldline which she regards as simultaneous with Billy's turning around. As before, $(v\tau, \tau)$ are the coordinates of event F, where v is Billy's speed relative to Alice. Let H be the event at which the flash of light is emitted which reaches Billy just as he turns around. We can deduce the coordinates of event H from the fact that the line HF is null. That means that the t-coordinates of H and F differ by the same amount as the x-coordinates. Since the x-coordinates differ by $v\tau$, it must be that the t-coordinate of event H is $v\tau$ less than the t-coordinate of F. It is now easy to see that H must have coordinates $(0, \tau - v\tau)$. Suppose Alice emits n flashes between events O and H. Billy then receives those n flashes between events O and F. Alice is emitting flashes at a rate of n per timespan of $\tau(1-v)$ seconds. Let f_A be Alice's emitting frequency, so $f_A = \frac{n}{\tau(1-v)}$. Since $\overline{OF} = \tau\sqrt{1-v^2}$, Billy is receiving flashes at a rate of n flashes per timespan of $\tau\sqrt{1-v^2}$ seconds. Let f_B be Billy's receiving frequency, so $f_B = \frac{n}{\tau\sqrt{1-v^2}}$. Using $1 - v^2 = (1+v)(1-v)$, you can verify that

$$f_B = f_A \sqrt{\frac{1-v}{1+v}} \qquad (8.5)$$

This answers our question: If alice emits 1 flash per second on her clock, Billy receives them at a rate of $\sqrt{\frac{1-v}{1+v}}$ flashes per second on his clock while he is outward bound.

The boxed formula is called the *relativistic Doppler shift*. Because $\sqrt{\frac{1-v}{1+v}}$ is less than 1, Billy perceives a frequency which is less than that emitted by Alice while he is outward bound. A similar Doppler shift occurs in sound waves when you are moving away from a siren: the frequency (pitch) of the sound is lowered.

The Twin Effect

> *Remark*: The relativistic Doppler shift is the primary means for converting telescope observations into the current model of the universe in which the galaxies are expanding away from each other due to a cosmic explosion (the big bang) which occurred some 10 or 20 billion years ago. Astronomers, by studying the light spectrum of a distant object, identifying particular lines, and measuring how much the frequencies of those lines differ from what they would be if produced in an Earth-based laboratory, can infer from that Doppler shift what the velocity of the object is relative to Earth. It is found that things farther away are moving away from us faster than nearby objects, and the speed is directly proportional to the distance from Earth. This observed correlation of distance with speed of recession yields the picture of an expanding universe. This established correlation can also be turned around to infer the distance of an object merely by measuring the Doppler shift of its spectrum. In this way astronomers deduce the distances of galaxies and quasars which are billions of light years from us.

While Billy is on his return journey, he also sees a Doppler shift in Alice's heartbeat frequency. But now the frequency is raised rather than lowered. (If you are *approaching* a siren the pitch of the sound is raised.) The formula which relates the frequencies is

$$f_B = f_A \sqrt{\frac{1+v}{1-v}} \tag{8.6}$$

> *Verification*: Suppose Alice emits m flashes in the time between events H and G. Since $\overline{HF} = \tau + \tau v$ seconds, her emitting frequency is $f_A = \frac{m}{\tau(1+v)}$. Billy receives the m flashes in the timespan $\overline{FG} = \tau\sqrt{1-v^2}$. So $f_B = \frac{m}{\tau\sqrt{1-v^2}}$. By comparing these expressions for f_A and f_B you'll find that $f_B = f_A\sqrt{\frac{1+v}{1-v}}$.

Now let's see how Billy watches Alice aging. His own age is easy – he just looks at his clock. Suppose $\hat{\tau}$ seconds is what his clock shows at event F, so $2\hat{\tau}$ is what it shows at G (since $\overline{OF} = \overline{FG}$). Between events O and F he receives flashes from Alice at a rate of $\sqrt{\frac{1-v}{1+v}}$ per second. Up to event F, therefore, Billy has observed $\hat{\tau}\sqrt{\frac{1-v}{1+v}}$ heartbeats of Alice's. Between events

F and G he receives the flashes at a rate of $\sqrt{\frac{1+v}{1-v}}$ for a total of $\hat{\tau}\sqrt{\frac{1+v}{1-v}}$ flashes between F and G. During his entire journey from O to G, therefore, he observes $\hat{\tau}\sqrt{\frac{1-v}{1+v}} + \hat{\tau}\sqrt{\frac{1+v}{1-v}}$ flashes, and you can simplify this expression to $\frac{2\hat{\tau}}{\sqrt{1-v^2}}$. This is Alice's age in seconds at event G, and comparing this to Billy's age of $2\hat{\tau}$ seconds, you see that Billy is younger by a factor of $\sqrt{1-v^2}$. And Billy has here done the reckoning.

Let's make sure we get the same result when Alice is observing heartbeat flashes from Billy. Refer to figure 8.13 Alice (if she is still living) can look at

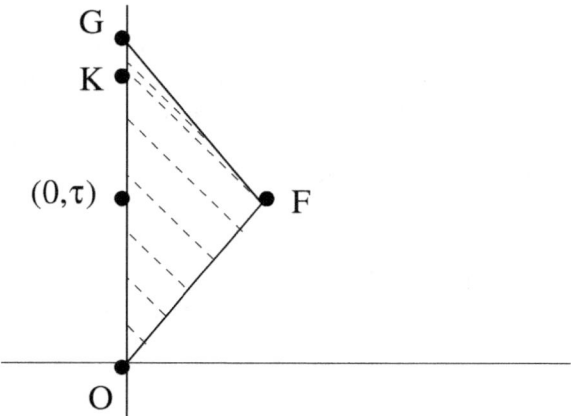

Figure 8.13:

her clock at G. Suppose it shows 2τ seconds. Then event F has coordinates $(v\tau, \tau)$. Let K be the event at which Alice receives the light flash from event F. In order for F and K to be null separated, you can see that K must have coordinates $(0, \tau(1+v))$. Between events O and K Alice receives flashes at a rate of $\sqrt{\frac{1-v}{1+v}}$. This is the Doppler shift formula, with Billy now emitting at 1 flash per second and the receiver Alice separating from Billy at speed v.

> *Note*: We can of course derive the formula again: p flashes in time $\overline{OF}$ yield an emitting frequency $f_E = \frac{p}{\tau\sqrt{1-v^2}}$, while Alice receives the p flashes in time $\overline{OK} = \tau(1+v)$ for a receiving frequency of $f_R = \frac{p}{\tau(1+v)}$. Comparing f_R to f_E, you'll find that $f_R = f_E\sqrt{\frac{1-v}{1+v}}$.

Between events O and K, therefore, Alice receives a total of $\tau(1+v)\sqrt{\frac{1-v}{1+v}}$ flashes from Billy, and this simplifies to $\tau\sqrt{1-v^2}$. After event K, the flashes arrive at a rate of $\sqrt{\frac{1+v}{1-v}}$ since the flashes are coming from an *approaching* source moving with speed v. (Check this formula for yourself, if you like.) In the timespan $\tau(1-v)$ between K and G, Alice receives $\tau(1-v)\sqrt{\frac{1+v}{1-v}}$ flashes from Billy, and this simplifies to $\tau\sqrt{1-v^2}$. Alice therefore sees a total of $\tau\sqrt{1-v^2}+\tau\sqrt{1-v^2}$, or $2\tau\sqrt{1-v^2}$, heartbeats announced by Billy during his journey. Notice that she too reckons that Billy is younger by a factor of $\sqrt{1-v^2}$ when they get together again at event G.

There is no paradox. Everyone agrees that Billy is younger than Alice by a factor of $\sqrt{1-v^2}$ when he returns at event G. Alice is always an unaccelerated person, and so she could simply apply formula 7.1 for time dilation to arrive at the correct conclusion. Because Billy reverses his motion, he needs to use more care in his computations.

Let me emphasize yet again how the twins' lives are unsymmetrical by virtue of Billy's change in motion. Each twin receives heartbeat flashes from the other at a rate of $\sqrt{\frac{1-v}{1+v}}$ for a while, and then at a rate of $\sqrt{\frac{1+v}{1-v}}$ until they meet again. Billy starts collecting Alice's flashes at the higher rate as soon as he turns around. Alice, on the other hand, does not see Billy's rate increase until light has had time to travel from the turn-around point back to Alice. (Compare figures 8.11 and 8.13.)

The twin effect is obviously of great significance to space travelers if they can achieve speeds close to the speed of light. A worldline like Billy's (figure 8.14) is an idealization. The corners of his worldline cannot be achieved with finite acceleration. On a sufficiently long space flight, however, moderate accelerations can cause a space traveler's worldline to approximate the appearance of Billy's (figure 8.15).

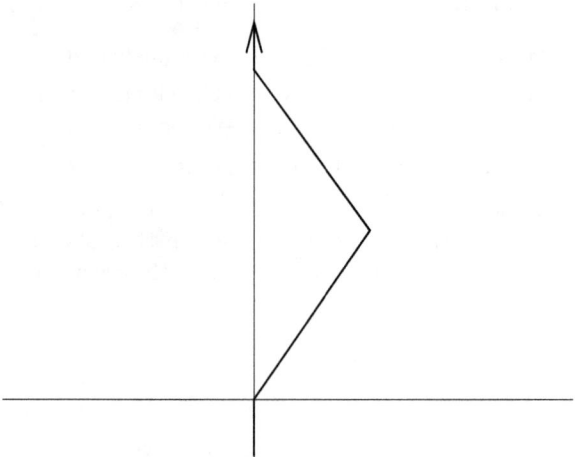

Figure 8.14:

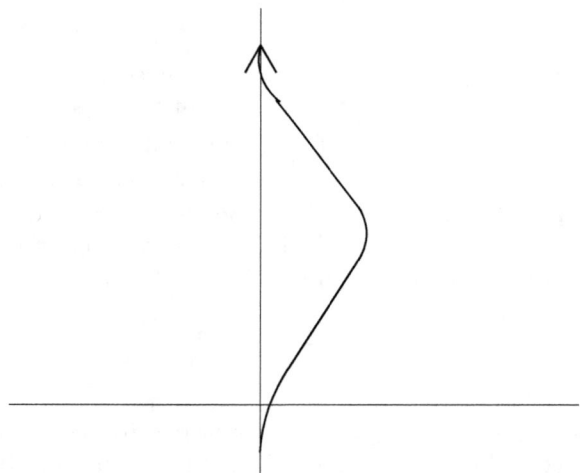

Figure 8.15:

Remark: Suppose a space person is willing to put up with an acceleration equal to the acceleration of gravity at the Earth's surface, so he "weighs" the same as on Earth. And suppose the direction of acceleration is reversed when halfway to the destination so the traveler slows down and comes to rest at the destination; and the return trip is made the same way with acceleration of constant magnitude. The space person's worldline would look like figure 8.16. A journey to and from a place 50,000 light years away (e.g. across our galaxy and back) takes a proper time of about 45 years[1] for such a space person. A journey to and from the Andromeda galaxy, about 2 million light years away, would take the traveler about 61 years by his clock. The vastly greater distance to Andromeda is traveled at speeds so close to that of light that not much extra proper time is required for the much longer journey.

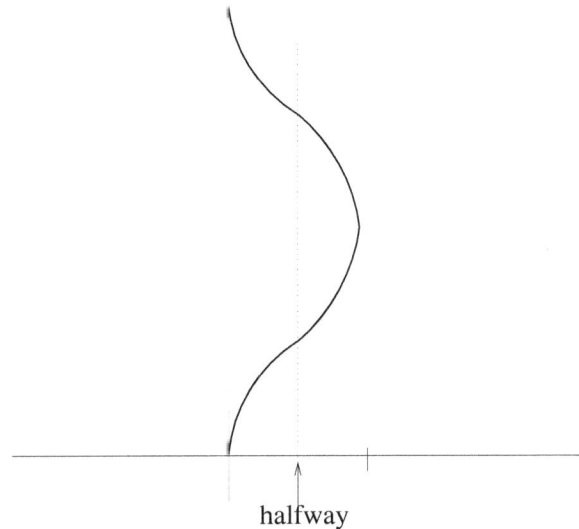

halfway

Figure 8.16:

Chapter 9

LORENTZ CONTRACTION

A moving object is shorter than the same object at rest. This effect is called Lorentz contraction.[1] By "a moving object" I mean an object which is moving relative to the person who measures its length. (The object is shortened only along the direction of relative motion, the other dimensions of the object being unaffected.)

A spacetime diagram will help you see why the length of an object depends upon the relative speed of the person making the measurement. But first let me try to interest you in a little puzzle.

Consider, if you will, a pole vaulter running with her pole at nearly the speed of light through a car garage. Suppose the pole is a little bit longer than the garage when they are compared with no relative motion. As the pole vaulter runs through the garage at high speed, the garage owner is able to trap the Lorentz-contracted pole in the garage by closing the front and back doors while the shortened pole is inside. Or so he thinks. From the point of view of the pole vaulter, the garage is moving at high speed toward her. The athlete measures the pole to be its normal length (since it is not moving relative to her), but she measures the moving garage to be shortened to a length very much shorter than the length of the pole. From the vaulter's point of view it is absurd to suppose that the pole is ever contained within the shortened garage. How can one reconcile these two points of view? Does the garage owner succeed in trapping the pole or doesn't he? I'll return to this puzzle after some general remarks about

---

[1] Lorentz, H.A., Amst. Verh. Akad. v. Wet. 1, 74 (1892). Fitzgerald, G.F., see Lodge, O., London Transact. (A) 184, 749 (1893).

Lorentz contraction.

Up to now I have drawn simply a line or a curve in spacetime to indicate the history of a person or object. At any instant, such an object or person would exist only at one point in space. But now we are interested in an object (like the pole vaulter's pole) whose spatial extent is the center of attention. A more detailed diagram is needed. Figure 9.1 displays the history of an extended object. The lines OG and EH are the worldlines of

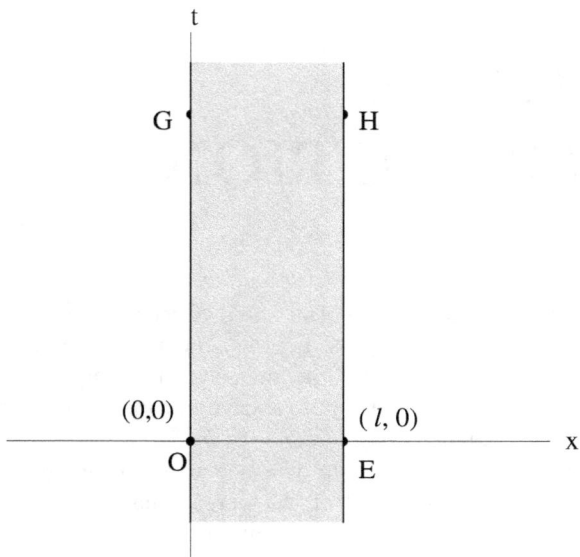

Figure 9.1:

the two ends of the object. The worldlines of all the intermediate parts of the object make up the shaded region of the figure.

If your worldline is the t-axis in the above diagram, then you are at one end of the object with no motion relative to it. You would determine the length of the object to be the distance $\overline{OE}$. Let's say event O has coordinates $(0,0)$ and E has coordinates $(l,0)$. Here l is the distance $\overline{OE}$, which is the length of the object as measured by you.

Let's now imagine some rocket moving with speed v past the object, and suppose events O and H are on the rocket pilot's worldline. Included in figure 9.2 is the line through event O which is orthogonal to the rocket's worldline. Event F is where this line meets the worldline EH. The rocket pilot regards all the events on the line OF as simultaneous with event O. From her point of view, one end of the object is at event O just when the

Lorentz Contraction

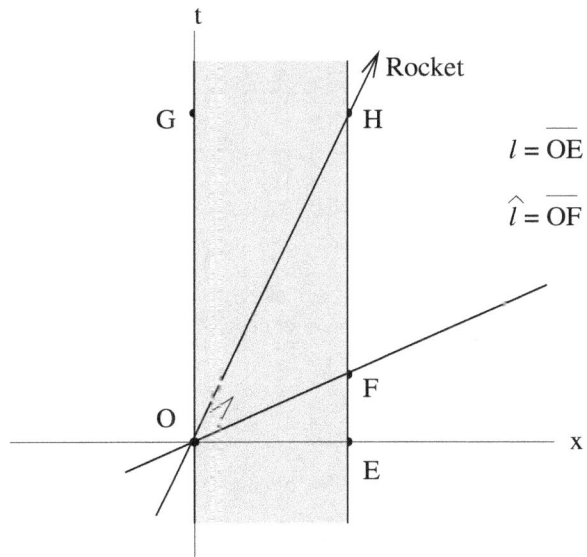

Figure 9.2:

other end is at event F. The distance $\overline{OF}$ is what she would measure as the length of the object (by radar, for example). I'll denote this distance $\overline{OF}$ by $\hat{l}$.

You measure the object's length to be l, and the object is not moving relative to you. The rocket pilot, moving at speed v relative to the object, determines its length to be $\hat{l}$. Let's find the quantitative relation between l and $\hat{l}$. In accordance with the techniques of chapter 4, the rocket's worldline OH has equation

$$t = \frac{1}{v}x \qquad (9.1)$$

and the equation for the orthogonal line OF is

$$t = vx \qquad (9.2)$$

The x-coordinate of event F is l, because that end of the object remains at $x = l$ for all values of t. Since the coordinates of event F must satisfy equation 9.2 for the line OF, it must be that the t-coordinate of F is vl. Using the coordinates for O and F, you'll find the Minkowski distance $\overline{OF}$ to be $l\sqrt{1-v^2}$. This is the length $\hat{l}$ which the pilot measures for the object; so

$$\hat{l} = l\sqrt{1 - v^2} \qquad (9.3)$$

This boxed equation is the quantitative statement of Lorentz contraction. Compared to its length when motionless, the length of an object is shorter by the factor $\sqrt{1-v^2}$ if it has speed v relative to the person measuring it.

Remark: It is worthwhile to check this result using a reference system in which the rocket pilot is stationary. See figure 9.3. The equation of the line OG is now

$$\hat{t} = -\frac{1}{v}\hat{x} \qquad (9.4)$$

and the orthogonal line OE has equation

$$\hat{t} = -v\hat{x} \qquad (9.5)$$

Suppose the rocket pilot measures the object's length to be $\hat{l}$ lightseconds. Then the $(\hat{x}, \hat{t})$ coordinates of event F are $(\hat{l}, 0)$. We want to determine what length you measure if you are moving with the object. That length is the distance $\overline{OE}$ since the ends of the object would simultaneously be at events O and E according to you. To compute the distance $\overline{OE}$ we need to know the coordinates of event E in the $(\hat{x}, \hat{t})$ system. Event E is the unique event which is on the line OE as well as the line EH. The slope of the line EH is $-\frac{1}{v}$ since it is parallel to the line OG. Knowing that event F, with coordinates $(\hat{l}, 0)$, is on line EH, you can check that the line must have equation

$$\hat{t} = -\frac{1}{v}\hat{x} + \frac{1}{v}\hat{l} \qquad (9.6)$$

The coordinates of E must satisfy this equation and also 9.5. Solving these two equations together gives the coordinates of event E:

$$\hat{x} = \frac{\hat{l}}{1-v^2}, \quad \hat{t} = -\frac{v\hat{l}}{1-v^2}$$

Using these coordinates of E, you can verify that the spacelike distance $\overline{OE}$ is $\frac{\hat{l}}{\sqrt{1-v^2}}$. As before, let l denote $\overline{OE}$, the length of the object at rest. Then l and $\hat{l}$ are related as in the boxed formula for Lorentz contraction.

Lorentz Contraction

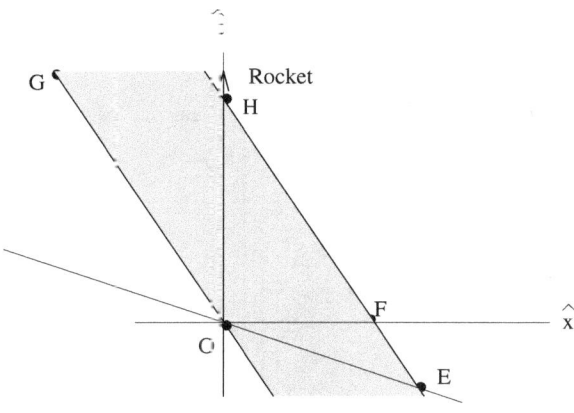

Figure 9.3:

Okay. Let's get back to the pole vaulter puzzle. We can imagine a garage with doors at each end and a pole which is slightly longer than the garage when they are compared with no relative motion. Now imagine the pole vaulter moving with the pole just fast enough that the Lorentz-contracted pole is the same length as the garage. In the reference system of the garage owner, the spacetime diagram of the moving pole looks like figure 9.4. The lines CH and FK are the wordlines of the two ends of the garage. Events O and F are events at which the worldlines of the pole ends meet the worldlines of the garage ends. By closing the doors at events O and F, the garage owner can trap the pole inside the garage.

On the other hand, the pole vaulter is justified in regarding herself (and the pole) as stationary and the garage as moving. Let's see what the spacetime diagram looks like in her reference system. The ends of the pole have worldlines parallel to the t-axis since they are stationary from her point of view. I'll sketch the worldline OH of one end of the garage, and also the orthogonal line OF (figure 9.6). The worldline of the garage's other end should also be drawn in the figure. That worldline must be parallel to the line OH since the two garage ends move at the same speed. Moreover, the line must contain event F because of the way event F has been defined. In the pole vaulter's reference system, the moving garage therefore looks like figure 9.7. Notice that the pole is never completely contained in the garage. At any time t, at least one end of the pole is outside of the moving garage. The pole vaulter therefore insists that the pole cannot be trapped.

Both points of view are correct. Either reference system may be used. The problem is this: What is meant by the pole being trapped? Suppose the garage owner decides to trap the pole forcefully, taking no chances on

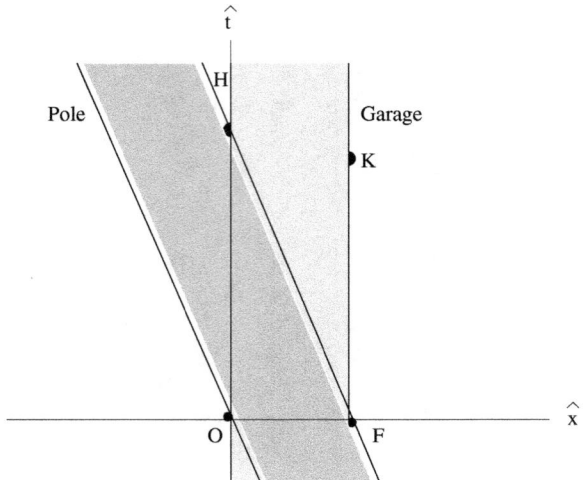

Figure 9.4:

Figure 9.5:

Lorentz Contraction

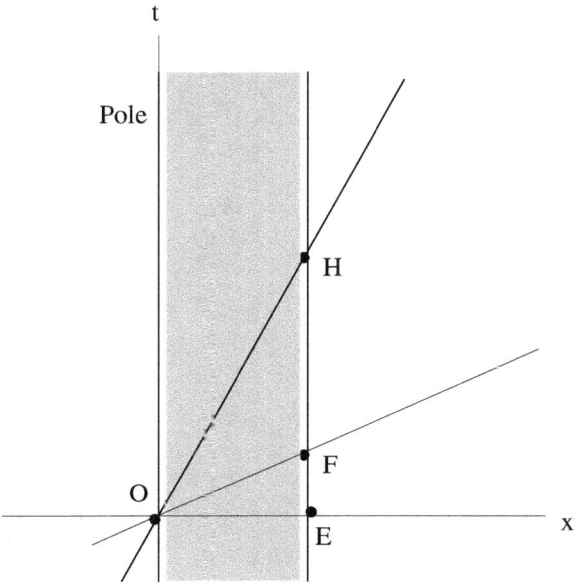

Figure 9.6:

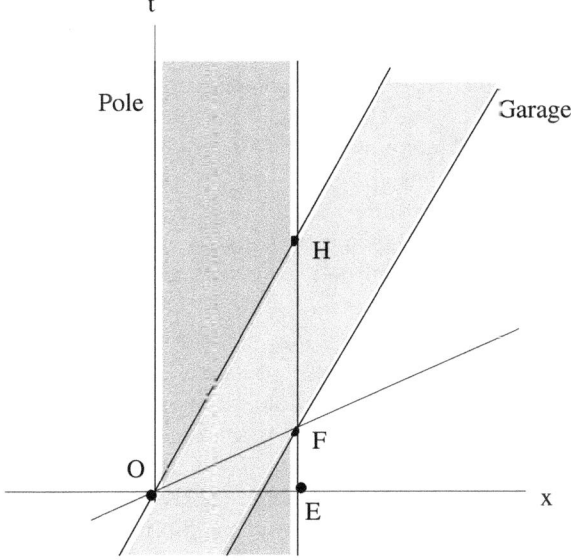

Figure 9.7:

it breaking out either end. So at time $\hat{t} = 0$ in figure 9.4 he staples the pole to the wall of the garage while the pole is inside. He simultaneously staples every point of the pole to the side wall, thereby stopping all points of the pole at once. Since the pole stops, the diagram 9.4 must be changed to figure 9.8.

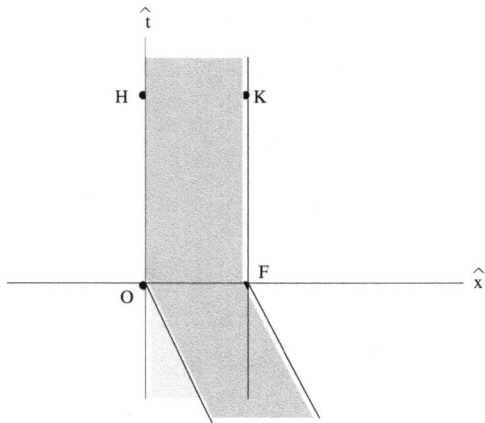

Figure 9.8:

In the other reference system, the history of the pole must be changed to look like figure 9.9. (Worldlines of some intermediate points on the pole have been included.) Note that, from this point of view, the points of the pole are not jolted into motion with the garage all at the same time. The vaulter would say that the pole was compressed from length $\overline{OE}$ to $\overline{CF}$ by this stapling procedure as the pole was set into motion with the garage. (Events O and F are simultaneous according to the garage owner, but C is the event which is regarded as simultaneous with F in the vaulter's reference system of figure 9.9.)

Although in figure 9.8 the pole may appear to be a rigid object to the garage owner, the stationary observer of figure 9.9, watching the same stapling of the pole, would say that the pole is not rigid – that it is compressed while it is being trapped. The moral to be gleaned from this is that one cannot talk sensibly about rigid objects. By a rigid object, one would *want* to mean an object whose ends always change their states of motion simultaneously. Two events which are simultaneous for one person, however, will not be simultaneous for somebody else in relative motion. Therefore, a jolted object which appears rigid to one person will not be regarded as rigid by somebody moving relative to the first person.

Lorentz Contraction

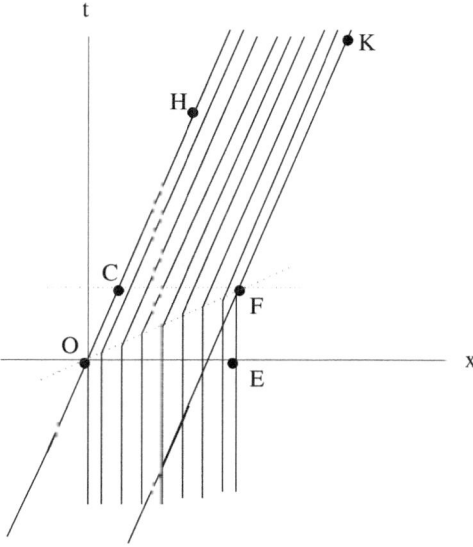

Figure 9.9:

In summary of the pole vaulter puzzle: Provided the garage owner does not attempt to stop the pole, the pole vaulter and the garage owner politely disagree because they have different ideas about which events are simultaneous. The garage owner is smug in his knowledge that he could trap the pole; the pole vaulter is confident that at any one time at least one end of the pole is safely outside the garage. On the other hand, the garage owner may prove his point by brutally stapling the pole against the wall all at once and verifying that both ends of the pole are indeed within the garage. The vaulter would regard that same stapling as poor sportsmanship, though, since from her point of view the staples are not driven in simultaneously. Indeed, the vaulter would say that the leading end is fixed first, and the pole is allowed to compress as the staples are installed, until the pole is as short as the garage by the time the following end is inside and stapled.

Recreation: Let k denote the length of a garage (when stationary), and let l be the length of a pole (when stationary). Suppose l is greater than k. Imagine a pole vaulter carrying the pole at speed

$$v = \sqrt{1 - (\frac{k}{l})^2}$$

through the garage. Verify that the garage owner would measure the garage and pole to be the same length. What are the lengths of the pole and garage as measured by the pole vaulter? (Answer: The pole length is l; the garage length is $k \cdot \frac{k}{l}$.)

Chapter 10

GENERAL RELATIVITY

In chapter one, I introduced 4-dimensional spacetime as a stack of 3-dimensional spaces, one copy of space for each instant of time; with that picture, the history of an object is simply a curve in spacetime. By itself, that notion of spacetime does not incorporate the ideas of relativity, and such 4-dimensional spacetime pictures have been available for centuries. The relativity phenomena of the preceding chapters stem from the use of the Minkowski interval formula for computing separations and distances between events.

Einstein recognized that the Minkowski spacetime geometry accurately describes the real world if one adopts the two rules (of chapter 2) along with the geometry: (1) the Minkowski null lines are the possible worldlines for photons of light. (2) If events E and F are timelike-separated, an accurate clock will measure a timespan equal to the separation $\overline{EF}$, provided the clock's worldline is the straight line segment from E to F. All the results which follow from these two rules in Minkowski spacetime are the domain of *special relativity*.

What is special about special relativity? From a physicist's perspective it is special because no effects of gravitation are included. Mathematically it is special because the Minkowski spacetime geometry is very special.

To appreciate the special properties of Minkowski spacetime one should examine other 4-dimensional geometries which can be meaningfully interpreted as spacetimes. Fortunately, the Euclidean plane is special among 2-dimensional spaces in the same way that Minkowski spacetime is special among 4-dimensional spacetimes. So it is enlightening to examine a variety of 2-dimensional geometries.

Consider, for example, the surface of the Earth, which I shall regard as a perfect sphere. Points of the surface may be labeled by coordinates of longitude and latitude. There are no straight lines within the surface to connect points, and the distance between two points depends upon the path chosen for measuring the distance. For example, a boat sailing a great circle route from New York to Lisbon will sail a shorter distance than a boat which sails at constant latitude.

> *Note*: *Great circles* on a sphere are the curves which you get in the following way: Pick any plane in 3-dimensional space which includes the sphere's center. Where the plane meets the sphere is a great circle (figure 10.1).

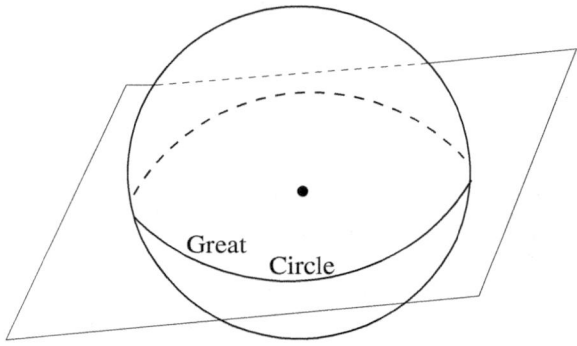

Figure 10.1:

As another example, consider the 2-dimensional surface of a peanut. Pairs of points do not have straight lines (in the surface) joining them, so one cannot talk about the distance between two points without first specifying a path along which to measure the distance. The same is true of the surface of a saddle, the surface of an automobile fender, jelly bean, etc. Among the possible 2-dimensional spaces, the plane is special in having a line segment joining any pair of points and a formula which gives the straight line distance between points.

In Minkowski spacetime any pair of events has a unique line segment joining them, and there is a formula (Minkowski's) which gives the interval along the line segment. These properties are not shared by the other spacetime geometries which are considered in *general relativity*. Minkowski spacetime is special.

More general spacetime geometries without straight lines are needed to account for the gravitational influences of massive objects. For example, there is a spacetime geometry which would be appropriate for a single star in an otherwise empty universe. A somewhat different geometry would prevail if two stars were orbiting around each other. The actual universe seems to be uniformly filled with galaxies, and so a model of the entire universe is a spacetime geometry which results from a uniform spatial distribution of matter at all times. Exactly how gravitating objects govern spacetime geometry is encoded in Einstein's field equation.[1] That is a technical subject which I shall have to omit from this discussion. The important thing is to realize that there *is* that relationship between gravitating objects and spacetime geometry. Minkowski spacetime is a spacetime with no gravitating objects. It is flat, with straight lines in all directions through all events. In the presence of gravitating objects, on the other hand, spacetime is curved and the straight lines are not available.

Since we live on the surface of a massive planet near a massive star within a galaxy of stars in a universe filled with such galaxies, it must be admitted that we do not live in Minkowski spacetime. It is nevertheless true that Minkowski spacetime is an excellent approximation to our actual geometry for most scientific purposes. There are only three situations in which one must consider the actual non-Minkowski geometry: (1) Spacetime curvature affects experiments and observations of very large scale. For example, the geometry surrounding the sun is a curved spacetime. Bending of light rays in this geometry can be detected because the light travels millions of miles after it is deflected and before it is observed with telescopes. (2) Even over small portions of spacetime, the deviations from Minkowski spacetime can be noticed if extremely precise measurements are made. For example, due to spacetime curvature caused by the mass of the Earth, Alice would be expected to outlive her twin brother Billy if she lives on the ground floor of a skyscraper building and Billy lives in the penthouse. It takes exceedingly accurate atomic clocks to detect the tiny difference in clock rates due to such a small difference in height above the Earth's surface, but it can be measured.[2] (3) Where gravitational effects are very strong, spacetime curvature is significant even over small portions of spacetime. For example, if you were to try living in the vicinity of a collapsed star, you could sense the curvature in the form of stretching and compression of different parts of your body.

These three circumstances in which Minkowski spacetime is *not* sufficiently accurate may be illuminated by considering the problems of making flat maps of curved surfaces. There are three situations in which flat maps are inadequate, and these closely parallel the three situations in which Minkowski spacetime is inadequate: (1) When the whole Earth, or a large portion of it, is represented on a flat map, there are serious distortions.

---

[1] Einstein, A., Ann. d. Phys. 49, 769 (1916).
[2] Pound, R.V. and Rebka, G.A., Phys. Rev. Lett. 3 439 (1959).

The effects of curvature are significant over large distances. (2) Even over small distances the effects of curvature can be noticed if extremely accurate measurements are made. A flat map of a city is accurate for most purposes, but distortions due to the Earth's curvature would show up with the use of accurate surveying equipment. (3) If an ordinary city were built on a small moon or asteroid, the city might cover a significant part of the sphere. In that case, even the taxi driver's odometer would reveal the distortions in his flat map of the city.

The preceding chapters of this book have concerned the geometry of Minkowski spacetime, and for most scientific purposes that geometry is suitable. In other words, we can use a Minkowski spacetime map for our neighborhood of the universe, and it will be accurate enough. The actual geometry we live in, however, must be quite different in its large-scale structure. To include the effects of gravitation, it is necessary to work with curved spacetimes, so let's see if we can build up some intuition about such geometries.

To begin with, recall our examples of curved 2-dimensional spaces: the Earth's surface, the surface of a jelly bean, a peanut shell, a saddle, and an automobile fender. In each case one can imagine the space as a surface within good old 3-dimensional Euclidean geometry. You should try, though, to forget about the 3-dimensional space in which the surfaces are embedded and consider *only* the points of the 2-dimensional space. Let's consider the Earth's surface in particular. Try to forget about the interior of the Earth and the space outside; consider just the surface. You can still tell that it is a curved space. Any triangle has more than 180 degrees as the sum of its angles, instead of exactly 180 degrees which it would have if it were in a plane. For example, the triangle in figure 10.2 has three 90° angles. (The sides of the "triangle" in this figure are not straight lines since there are no straight lines on a sphere. The triangle's sides are portions of great circles. Any path of shortest length between two points on a sphere is a portion of a great circle. For this reason, segments of great circles are appropriate as the sides of a triangle on a sphere. Such segments are as straight as they can be.)

By measuring angles and distances, a person confined to a 2-dimensional space (like the world's surface) can determine its shape. If he is placed in a curved 3-dimensional space, the person can still determine the geometry by the same types of measurements. Three-dimensional curved spaces are more difficult for us to imagine, but let's study one anyway.

I'd like you to imagine the 3-dimensional surface of a 4-dimensional ball. Consider first a 4-dimensional Euclidean space. Let the points of the space be labeled by the four coordinates w, x, y, z. Even though you can't visualize this 4-dimensional space, you can recognize that the point

General Relativity

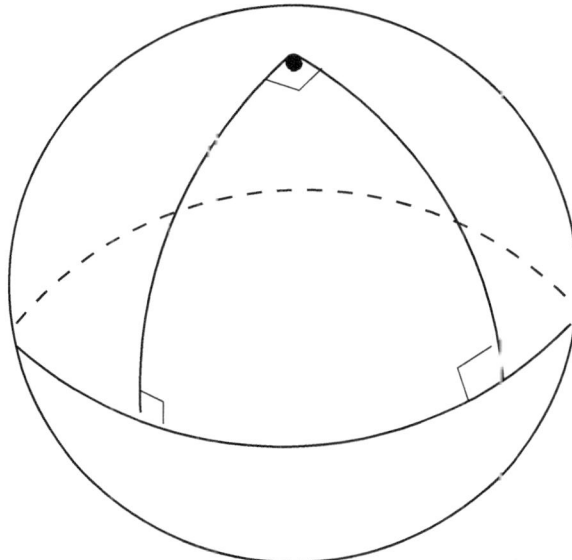

Figure 10.2:

(3,2,3,4) is a point different from (1,4,5,8), for example. The distance between (w_1, x_1, y_1, z_1) and (w_2, x_2, y_2, z_2) is to be

$$\mathcal{D} = \sqrt{(w_2 - w_1)^2 + (x_2 - x_1)^2 + (y_2 - y_1)^2 + (z_2 - z_1)^2} \qquad (10.1)$$

All the points of this 4-dimensional space which are less than r units away from the origin (0,0,0,0) form a 4-dimensional ball of radius r. The points which are at a distance of exactly r units from the origin constitute the surface of that ball. The points on that 3-dimensional surface are the points (w, x, y, z) which satisfy the equation

$$\sqrt{w^2 + x^2 + y^2 + z^2} = r \qquad (10.2)$$

Now suppose you are placed inside this 3-dimensional space (i.e. inside the *surface* of the 4-dimensional ball) and you are allowed to explore it. Since it is 3-dimensional, you feel quite at home. If the radius is large, you might not realize that it is other than good old 3-dimensional Euclidean space, just as most humans thought the Earth is flat until about 500 years ago. Suppose you set out and try to fly in a straight line. As on the Earth's surface, there is no straight line. The best you can do is to fly on a "great circle." (Your position and initial direction define a line, which defines a unique plane in the 4-dimensional Euclidean space that also contains the origin (0,0,0,0). The points of the 3-dimensional sphere which also belong to this plane constitute the "great circle" for your flight.) Following along the "great circle," you would eventually return to your starting point. Such exercises would soon convince you that your 3-dimensional space is in fact curved, even though you might never consider the 4-dimensional Euclidean space I introduced to facilitate the description of this particular 3-dimensional space. You could determine all the properties of the space by measuring angles and distances within the 3-dimensional space.

An important model of the universe can be built up from such 3-dimensional spheres. One imagines that, at any instant of time, the universe is a 3-sphere (as above) of radius r, i.e. a 3-dimensional space that could be embedded in 4-dimensional Euclidean space as the surface of a ball. The radius r is increasing with time. The galaxies are assumed to be evenly distributed throughout the 3-sphere at each moment of time. As r increases, the galaxies get farther from each other because the 3-sphere is expanding. (It is analogous to a spotted balloon – a spotted 2-sphere – which is being inflated: as the radius of the balloon gets bigger, all the spots move away from each other in the surface.) In this model, the universe has a finite volume at any time, and yet it has no boundary! It is not known if the volume of the entire universe is finite or not. If it is finite, it has an extremely large radius that makes it appear to have no curvature at all. Some cosmologists contend that the universe experienced a powerful "inflation" immediately after the big bang, and the inflation made space into a 3-dimensional version of an inflated balloon that has undetectable curvature in its surface.

As in Minkowski spacetime, a curved spacetime has a null cone at each event. Instead of null *lines* through the event, we must now think of the null cone as consisting of null *directions* at each event. Directions inside the null cone are timelike and directions outside the null cone are spacelike. See figure 10.3. A curve is said to be null if its direction is null at each event of the curve. A timelike curve has a timelike direction at each of its events.

General Relativity

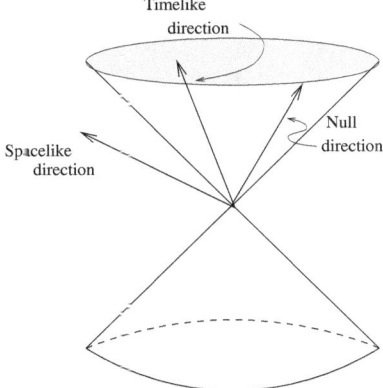

Figure 10.3:

Remark: A black hole is a region of spacetime from which no particles can emerge. It is black because no particles of light (photons) come away from it. It is a hole because stuff can fall into it. Figure 10.4 represents a black hole. The light cones are shown at a number of spacetime events. (Unlike my previous spacetime diagrams, null directions are not all tangent to 45° lines in this figure.) Notice that, near the middle of the picture, the light cones are tipped inward.

The cylinder in the picture represents the history of the black hole's "surface". Unlike the surface of a planet or star, there is no matter at the "surface" of a black hole. From events inside the "surface" you could not send signals to the outside, because all future-directed timelike or null curves point inward. In contrast, signals *can* escape from any event which is outside the "surface." The cylinder is called the "event horizon" of the black hole because observers on the outside of it can never see anything arriving from events inside.

Each horizontal slice of the cylinder looks like a circle in the picture. If I could accurately sketch the 4-dimensional spacetime and show all 3 spatial dimensions as well as time, then each slice would look like a sphere. You can think of each sphere as the black hole's "surface" at a particular instant of time.

Black holes are supposed to result from the collapse of some stars after they burn up all their nuclear fuel. There is also evidence for much larger black holes at the centers of many galaxies.

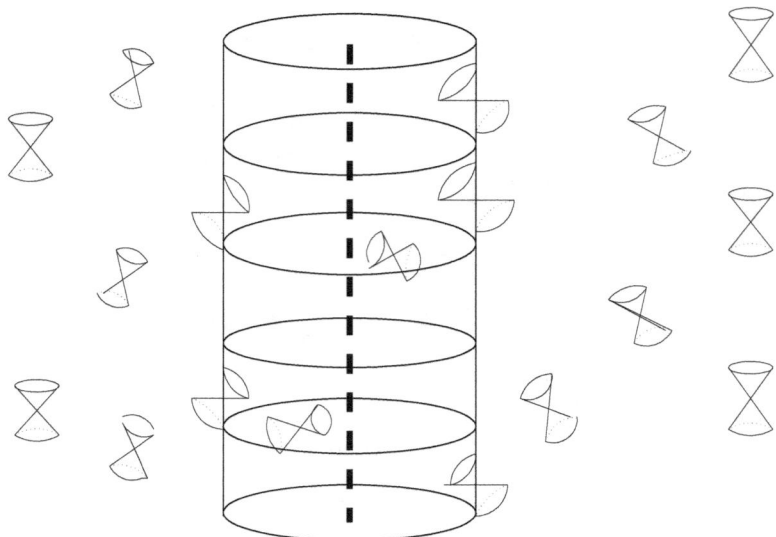

Figure 10.4:

Because spacetime geometry is not perceived directly by human beings, we need the aid of a few instructions in order to see how the geometry describes the real world: (1) The worldline of a photon is a null curve. (2) If the worldline of an accurate clock is a timelike curve from event E to event F, then the clock will record the length of the curve as the timespan between the events. These two assumptions are of the same form as the rules adopted for Minkowski spacetime. In general relativity a third assumption is necessary. It is called the *geodesic hypothesis*: (3) An object experiencing no acceleration will have a *geodesic* worldline. In a flat space, or flat spacetime, geodesics are the same as straight lines. In a curved geometry, the geodesics are as straight as they can be. On a sphere, for example, the geodesics are the great circles. In more complicated geometries the geodesics are not so easy to characterize. Any space or spacetime, however, has the following property: if a point is specified along with a direction at the point, then there is a unique geodesic which passes through the specified point in the specified direction.

An especially interesting example is the motion of the Earth in the spacetime geometry near the sun. One type of geodesic in that geometry is like a helix centered on the worldline of the sun. See figure 10.5 The Earth has a worline which is a geodesic of this type. By reading the spacetime diagram upward so as to produce a motion picture, the Earth is seen to go

General Relativity

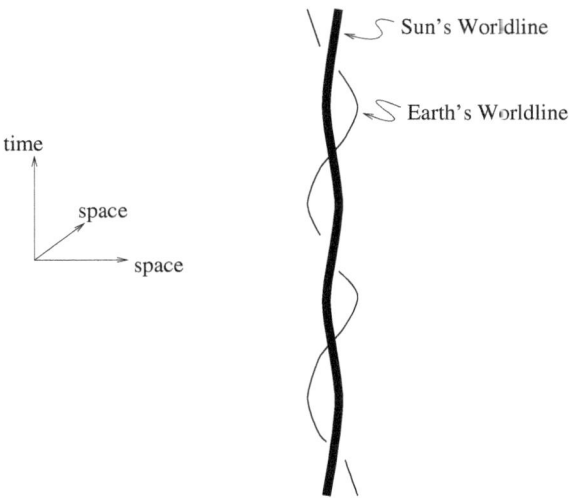

Figure 10.5:

round and round the sun.

Notice that the Earth's worldline is a geodesic – the Earth is not accelerating. It's worldline is as close to a straight line as it can be in the curved spacetime geometry which surrounds the sun. This contrasts sharply with Newton's picture of gravitation. In Newton's theory, an unaccelerated object would move along a straight line in space. The sun exerted a gravitational force on the Earth which caused the Earth to accelerate and thereby run circles around the sun instead of shooting off along a straight line. Einstein's version of gravitation is radically different. The Earth is embedded in a spacetime geometry which is somewhat different from Minkowski spacetime due to the presence of the sun and the planets (including Earth). The worldline of the Earth is a geodesic of this geometry, i.e. a generalized straight line. The Earth is not accelerating. There is no force exerted on the Earth; any such force would cause the Earth to accelerate and not have a geodesic worldline. In general relativity, you see, the whole idea of a "gravitational force" has been discarded. Objects like the sun and Earth influence each other by virtue of their effects on the spacetime geometry. Each object follows a geodesic of the geometry. There is no place for Newton's gravitational force in general relativity, so gravitation is conceptually quite different from Newton's idea of it.

Remark: Although Newton's theory is not conceptually correct, it is extremely useful because (1) for the everyday calculations which people do, general relativity gives almost precisely the same answers as Newton's theory, and (2) the Newtonian calculations are usually much easier to do.

There are, of course, non-gravitational forces which may act on an object. Those forces will cause the object to accelerate, i.e. the object feeling the force will *not* have a geodesic worldline. As you sit in your seat, you feel a force. The seat is pushing up on you, accelerating you away from a geodesic worldline. Imagine that there were no seat, no floor, no ground, and no air beneath you. That is , suppose there were an evacuated tunnel right through the Earth. (And, for simplicity, pretend that the Earth were not rotating.) Your geodesic worldline would take you through the center of the Earth to the opposite (anti-podal) point, then back, and you would continue to oscillate in that way (see figure 10.6). That would be your natural motion in the spacetime geometry of the Earth if your seat were

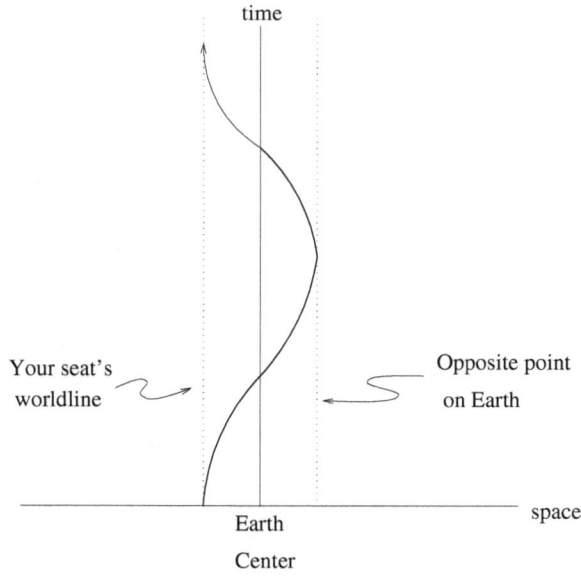

Figure 10.6:

suddenly pulled out from under you. You would feel no forces along that worldline. Your seat accelerates you away from the geodesic worldline, and you can feel the force of the acceleration.

Notice how neatly general relativity explains Galileo's observation that all objects fall at the same rate. Any object, when you let go of it, will start following a geodesic. The object simply follows the closest thing to a straight line. The objects which Galileo released all experienced the same geometry, and the geodesic worldlines were all the same.

That is the gist of relativity. Matter causes spacetime curvature, and non-accelerated objects enjoy geodesic worldlines in the curved spacetime. Gravitation is geometry – there is no gravitational force. Minkowski spacetime with its straight-line geodesics is the special spacetime with no gravitation. Its geometry is the essence of special relativity.

About the Author

After receiving a B.A. degree in physics from U.C. Berkeley (1968) and an M.S. degree from U.C San Diego (1969), Paul Sommers moved to Austin to study in the Relativity Center under the supervision of Alfred Schild at the University of Texas. He was strongly influenced by Martin Walker and Bob Geroch, and received his Ph.D. in 1973. He then worked for two years at Oxford as a postdoc with Roger Penrose, followed by two years at UNC in Chapel Hill with Jim York. In 1977 he became an assistant professor of mathematics at NC State University. He and his wife (Gayle Nicholson) moved aboard a 31-foot trimaran sailboat in 1980 and spent two years sailing and diving in the Bahamas and western Caribbean. Following that, they moved to Salt Lake City, where he developed an interest in cosmic ray research at the University of Utah. In 1995, he helped to design the Pierre Auger Cosmic Ray Observatory as a founding member of that international collaboration led by Jim Cronin. He became a full professor of physics at Penn State University in 2005 where he served as associate director of the Institute for Gravitation and the Cosmos led by Abhay Ashtekar. His research remained concentrated in the study of the highest energy cosmic rays, and he was co-spokesperson for the Pierre Auger Collaboration 2007-2010. He retired in 2013, and he and Gayle moved to Wilmington NC where they live on the shore of the Intracoastal Waterway overlooking Carolina Beach Inlet. Their daughters, Pacifica and Aleah, live in Boulder, CO.

www.ingramcontent.com/pod-product-compliance
Lightning Source LLC
Chambersburg PA
CBHW070146230526
45471CB00002B/533